# Needlework Designs from the American Indians

# Needlework Designs from the American Indians

## Traditional Patterns of the Southeastern Tribes

Anne Cheek Landsman

South Brunswick and New York: A. S. Barnes and Company
London: Thomas Yoseloff Ltd

A. S. Barnes and Co., Inc.
Cranbury, New Jersey 08512

Thomas Yoseloff Ltd
Magdalen House
136-148 Tooley Street
London SE1 2TT, England

**Library of Congress Cataloging in Publication Data**

Landsman, Anne Cheek, 1948–
    Needlework designs from the American Indian tribes of the Southeastern United States.

    Includes index.
    1. Canvas embroidery—Patterns. 2. Indians of North America—Southern States—Textile industry and fabrics. I. Title.
TT778.C3L27        746.4'4        76-10871
ISBN 0-498-01804-0

Printed in the United States of America

# CONTENTS

Acknowledgments     7

Introduction     9

1 A Short History of the Southeastern Tribes     13

2 The Stitches     18

3 Know Your Craft     28

4 The Designs     37

Index     160

# ACKNOWLEDGMENTS

I would like to thank the people who have given me encouragement and have helped in so many ways.

Jacob Ayyancovil
Melissa Bogle
Sterling Bogle
Book World, Inc.
Roberta K. Cheek
S. L. Cheek, Jr.
Mark Dittrick
Mrs. Irma Jones
Diane Kender
Rita Krever
Sonia Landsman
Wallis Mayers Needlework, Inc.

And to my husband, Joseph K. Landsman, without whose encouragement, patience, and love this book would not be possible.

Photographs by Mark Dittrick.

# INTRODUCTION

In recent years, a great deal of attention has been focused on all aspects of the society of the first Americans—the Indian. Bookshelves are crowded with accounts of the American Indian and his relationship to American history and culture; special issues and sections of many popular magazines are devoted to his craftsmanship, his artistic and social achievements; stores and outlets of all kinds display the crafts and wares being produced by tribal cooperatives and individual craftsmen.

There is a wealth of representative arts and crafts from many of the tribes whose names, once memorized from early history lessons, bring forth images of colorful handicrafts that now adorn our homes and ourselves. From the Southwest tribes—Navaho, Zuni, Hopi—come brilliant silver and turquoise jewelry, colorful weavings, sand paintings, intricate basketry, and pottery. From the Northwest tribes—Kwakiutl, Haida, Nootka, Tlingit—come the traditional totemic art, fine bone and woodcarvings, button blankets. The once great hunters of Great Plains tribes—Cheyenne, Sioux, Blackfeet, Crow, Mandan—and the more practical farmers of the Woodlands tribes—Iroquois, Seneca, Ojibway, Algonquin—now provide apparel and various household objects fashioned from leather, beadwork, birch bark, stone, and a variety of other traditional and natural elements.

Until recent years, however, there was little evidence in the craft world of the tribes of the Southeast—Cherokee, Creek, Choctaw, Chickasaw, and Seminole—but that is changing. Blackware pottery, river cane basketry, woodcarving, beadwork, wool and linen weaving, jewelry, appliqué, and, perhaps most notably, Seminole patchwork are just some examples of present-day crafts of the Southeastern Indian tribes.

My own heritage as a part-blood Alabama Cherokee stirred me to go further than preparing yet another needlepoint design book. In gathering and presenting the material on the following pages, I have tried to provide you with more of an insight into the source and symbolism of the designs that have been adapted for the needlepoint canvas. The brief backgrounds of the Southeastern tribes along with the information accompanying each design will, I hope, make each project more interesting and alive, and add to your knowledge and understanding of some of the peoples to whom we owe much of our own native American heritage.

# Needlework Designs from the American Indians

# 1
# A Short History of
# THE SOUTHEASTERN
# TRIBES

## The Cherokee

The Cherokee have always taken pride in their ability to succeed under adverse conditions. The ancient Cherokee were a warrior and hunter nation who, by 1700, controlled over seventy million acres of Kentucky, Tennessee, North and South Carolina, Virginia, Alabama, and Georgia. The vast Appalachian range was their domain until the colonial wars decimated both their holdings and their people. Their survival during this period was not just a miracle, but a tribute to their perseverance and wisdom. Instead of engaging in futile opposition, by the end of the American Revolution the Cherokee had assimilated into white society. They continued their upward climb even after their removal from their homeland in the Southeast to Eastern Oklahoma during the "Trail of Tears," an infamous journey in American and Indian history that took its toll of almost four-thousand of the fifteen-thousand Cherokee forced to make the journey.

In the years that followed, the tribe formed a constitutional government, invented a syllabary (accomplished in 1821 by Sequoyah), established a newspaper, opened their own free public school system, and created a prosperous farm and merchant economy. Since tribal customs and culture have always been uppermost in the life of the Cherokee, those traditions, even today, have not changed. *The New Cherokee Advocate,* the tribal newspaper, is printed in both English and Cherokee. The spirit of their ancestors is evident in the crafts being created by the Oklahoma tribe and the Eastern Band of Cherokee still residing in North Carolina.

The predominant color schemes that find their way into most of the

Cherokee work are both traditional and symbolic. Red is the color of the East and the most sacred of all colors of the ancients for it represents success and triumph. Blue, color of the North, is defeat and trouble, while black, color of the West, means death, not unlike its symbolism in most cultures. Finally, white, color of the South, represents peace and happiness. In addition, the balance of colors described on some of the following pages represents the basic earth colors of the Southeast.

The Cherokee tradition can be found most in their blackware pottery, which has gained attention for its beauty and grace. Along with their pottery, the Cherokee produce fine cane basketry, leatherwork, and weavings.

## The Creek

Like the Cherokee, the Creek were an organized and socially ordered tribe long before their contact with European civilization, even though religious superstition and magic played a major role in their society. The *Esaugeta Emissee* or Master of Breath was the great spirit who watched over the Creek. His domain also included many animals in places of honor such as the wolf and rattlesnake. Their preoccupation with the mystic provided charms, bad luck omens, and various mythical creatures. Among the most prominent were the great horned snakes and *tie* snakes, friendly creatures who often captured a Creek just to show him hospitality. As a result, the priests held positions of great reverence in the society. Festivals also played a major role in their culture, the most famous of which was the annual *boosketah* or *busk*, which was culminated by the Green Corn Dance. It was the Creek version of the end of the old and the beginning of the new year.

Unlike the Cherokee, however, the Creek did not fare well in the period following the colonization of the New World. From pre-Revolutionary days through the early 1800s, the Creek were forced to give up most of their holdings through defeats by the English and then by the Americans. Though they finally acceded to removal in 1832, the weight of all the abuses and fraudulent land practices resulted in the Creek War of 1836. It was after this final defeat that the Creek were moved to the Western territory, losing more than forty percent of their population along the way.

Being a warrior nation, their adjustment to reservation life was slow and painful. It was just prior to the Civil War that education finally took hold among the Creek. In the postwar period, internal struggles held back the tribe from unity and progress. Only when the leaders of the Creek nation were called upon to take the leadership among the Five Civilized Tribes were they able to assume some leadership among themselves. Finally in 1936, as a result of the Oklahoma Indian Welfare Act, the tribe began to establish a self-sufficient economy.

During their bloody and troubled history, the Creek remained true to their ancient social and cultural ways. The Green Corn Dance and Festival is still very much a part of modern tribal life, while in the craft fields, contemporary leatherwork reflects old expertise in shoulder bags, belts, and moccasins. Body tattoos, a warrior tradition, and other painted body ornamentation are now the source for a wealth of designs adapted for needlepoint.

## The Choctaw

Contrary to the history of most Indian tribes, the Choctaw story is not filled with bloodshed, violence, harsh poverty, or fabulous riches. Instead, the Choctaw were, and still are, a practical and peaceful people who quietly managed to meet the problems of their constantly changing environment.

Choctaw legend tells us that the two founding brothers, Chahta and Chicksa, left their people to search for their homeland. After crossing what is now the Mississippi, they took different paths. Chicksa settled soon after in Western Mississippi and founded the Chickasaw tribe while Chahta found a home in Central and Eastern Mississippi for the Choctaw tribe. Primarily a hunting and farming people, the were often the target for slavery and their history recounts many raids by their kinsmen, the Chickasaw. Years of strife saw their numbers decimated, their lands confiscated, and many of their people enslaved. By the 1830s, they were ready for removal and it did not take them long to establish a strong farm economy in their new environment. They began to branch out into industrial activity building saw mills, salt works, cotton mills, even obtaining a government contract to manufacture looms.

Economic growth was accompanied by social recovery. The Choctaw developed their own language utilizing English, and education became a watchword for all. As it did to other tribes, the Civil War set back their growth, but by the time the Oklahoma Indian Welfare Act was passed in the 1930s, the Choctaw were already on their way to reestablishing themselves as an industrious and ambitious people. Today there is a revived sense of heritage among the Choctaw people. Since the Choctaw had a strong farm economy by tradition, basketry and pottery still remain a fine art as in the past, along with beadwork, belts and sashes, wool and linen weaving, and leatherwork.

As an additional note, a number of Choctaw designs in this book have been taken from the belts of lacrosse players. The original game, called *baggataway*, is of Indian origin, probably Choctaw or some other tribe of Southeastern origin. The name comes from the French *la crosse*, meaning *cross*. The early French explorers, seeing the Indians at their game, likened the curved playing sticks to the many crosses used in their own religious ceremonies.

## The Chickasaw

In sharp contrast to the peaceful attitudes of the Choctaw, the Chickasaw were a strong and proud warrior nation, a direct cause of their small number today. From the first of DeSoto's expeditions through the American Revolution, the Chickasaw met every attempt at colonization with resistance. In fact, the opportunist warriors hired out as mercenaries preying upon other tribes to provide the settlers with slaves and booty. During the first one-hundred years of contact with European civilization, the Chickasaw had become so dependent upon the tools and customs of that culture that they lost their self-sufficiency and independence. Intermarriage reduced their pure-blood number drastically and, more important, buried much of their culture and heritage. The tribe was constantly beset by internal struggle between pure- and part-blood factions, warring for power. This division made the confiscation of their lands an easy task, resulting in their removal in 1832 to the Oklahoma territories. By the time removal of the Chickasaw was completed, fewer than six-thousand of this once great warrior race remained. In 1844, four-thousand of them resided in Choctaw territory, and it is only in the last three decades of this century that some unity has helped to reshape the Chickasaw as a nation. Still, only eight-thousand, most of whom are mixed blood, inhabit the Indian reservation in Oklahoma today.

If economic and social development has been long in coming for the Chickasaw, then cultural rebirth has been even more difficult. The remnants of a once proud and powerful people still lie buried somewhere in the unearthed recesses of the lower Mississippi Valley. Only a Chickasaw desire for resurrection of old and cherished traditions will bring back their ceremonies, customs, and crafts.

I have attempted, without success, to include some Chickasaw designs in this book. A myriad of sources were contacted: Southeastern States historical societies, Indian societies, anthropologists, museums, ethnology records from the federal government and various universities, and even the Chickasaw tribe itself. All of these efforts were rewarded by a single photograph of a lacrosse stick and ball from the dusty archives of the Smithsonian Institute in Washington, D.C. If any of my readers can furnish me with authentic Chickasaw material or have seen such material, I would appreciate the information.

## The Seminole

Of the five major Southeastern tribes, the Seminole are the most unusual. They do not boast a long history but are, in fact, an offshoot of the Creek of Georgia. In 1716, the Georgia Creeks, fearing reprisals as a result of the Indian wars and raids being waged in the Carolinas, chose to move south into Florida. Welcomed by the Spanish, who called them *Cimarrones* or wild ones, they managed to settle in the wild forests and swamps, which were of no consequence to the Spanish and, later, the English. In the desolation of the swamps, and with as little contact with the Spanish as possible, the Seminoles began to structure their society as they had done in Georgia—in well-ordered Creek fashion.

During the Creek War of 1814, many of their distant brothers fled south, finding sanctuary with the Seminole. As their numbers grew, so did friction between England and the United States, and, while England used the tribe for its own purposes, the United States waged war on the tribe in the First Seminole War of 1817, in order to bring about Spanish secession of the Florida territory. Though the federal government signed numerous treaties with the Seminole between 1823 and 1832, 1836 saw the attempt at forced removal of the tribe and the Second Seminole War, which lasted until 1842. Out of desperation, a major portion of the tribe went westward, while a small band moved deeper into the swamps. In 1855, the Third Seminole War was fought in an attempt to remove the remaining Seminole from Florida but, fighting on their own terms, and finally reduced to about three-hundred, they did not succumb. From this small band, the Florida Seminole tribe now stands at twenty-five-hundred who have fashioned a strong economy in relative obscurity in the dense Everglades and, as they proudly claim, have still not signed a treaty of peace with the United States.

In spite of Florida's growing population, the tribe still lives quietly in the far reaches of the swamp country. They are cattle and farm owners, hunters, fishermen, and guides, and they manufacture a small variety of craft items for a flourishing and curious tourist trade. Lacking in a long and rich background, the most notable Seminole craft is patchwork. It is not ancient but uniquely Seminole, and its variety of color and design has made it a major contribution to native American Indian handicrafts. Most of the Seminole designs in this book have been adapted from their patchwork with a few designs taken from some old and rare pottery and jewelry pieces. Though lacking in symbolism, the principal colors used by the Seminole in their patchwork are red, yellow, black, and white.

# 2
# THE STITCHES

Needlework is much more interesting when worked in a variety of stitches. This variety adds another dimension to the variety of colors suggested for the Southeastern Indian designs. Some of these stitches will also add texture to your work, since I believe a piece should *feel* as well *look* interesting.

When reading these stitch diagrams, remember that they are always larger than the canvas you are using. Even though the stitch looks tremendous on the diagram, it will be smaller and tighter when you work it. Every space between lines of a diagram corresponds to the space between threads on your canvas. For instance, if a stitch count on a diagram is over three spaces, it will be over three mesh squares on your canvas.

The diagrams are keyed by number. The first thing to remember is that the needle comes out of the canvas on the *odd* numbers and goes into the canvas on the *even* numbers. Second, remember to count the number of spaces indicated by the diagrams both *horizontally* and *vertically*. Keep these two instructions in mind and you will find each stitch easy to follow.

You will notice that there are two different sets of diagrams for the seven stitches used in this book. The first set is for right-handed needlepointers, the second for those of you who are left-handed and often neglected.

## Right-Handed Stitch Diagrams

*Continental Stitch:* The continental is worked from right to left. As you work you will be moving one hole to the left of the previous stitch. When you come to the end of a row, turn the canvas upside down *before* you start the next row.

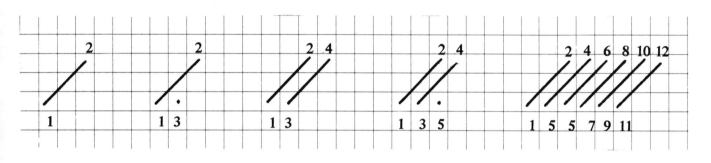

*Slanting Gobelin:* This is a large version of the continental stitch and especially good for background areas.

This is the slanting gobelin stitch for right corners.

This is the slanting gobelin stitch for left corners.

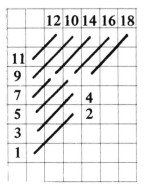

*St. Andrew Cross* and *St. George Cross:* These are accent stitches and should be worked last on any design that makes use of them. The St. Andrew cross looks especially good when worked alone in a small area such as the Seminole pillow described in the Seminole tribe section.

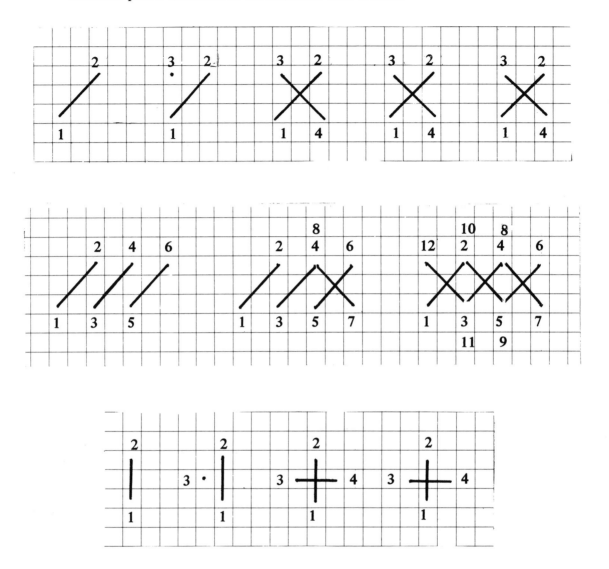

*Straight Gobelin:* This also makes a good background stitch. When worked over three mesh squares it resembles woven cloth and certain types of basketry weaving.

**Brick:** This is a series of straight gobelin stitches worked over three mesh squares and staggered to look like rows of bricks. Bricking is excellent for backgrounds or creating subtle textures.

**Flame:** This, too, is a series of straight gobelin stitches worked over three mesh squares in a step pattern to form a point. It can also be worked over as many mesh squares as you desire to form either a larger or smaller point.

*Cherokee Diamond:* This is a stitch I created especially for the designs of the Southeastern tribes. It is worked counterclockwise, with *each* stitch overlaying the one before it. This creates a "bump" effect that works nicely with the other stitch textures. Thought it may look fancy and complicated, it is easy and fun to work.

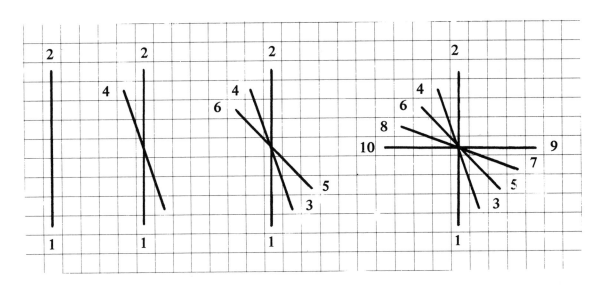

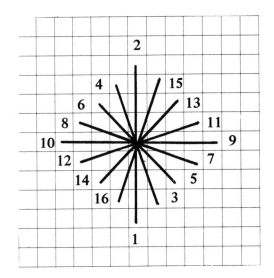

22

# Left-Handed Stitch Diagrams

*Continental Stitch:* The continental is worked from right to left. As you work you will be moving one hole to the left of the previous stitch. When you come to the end of a row, turn the canvas upside down *before* you start the next row. Two methods of working this stitch have been included. Choose the one that works easiest and most naturally for you.

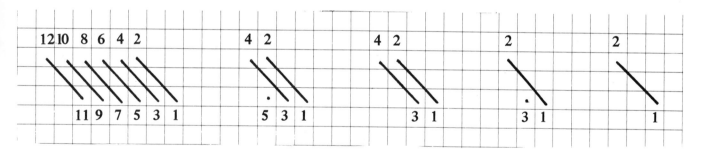

*Alternate Continental*

*Slanting Gobelin:* This is a large version of the continental stitch and especially good for background areas. Two methods of working this stitch have been included. Choose the one that works easiest and most naturally for you.

This is the slanting gobelin stitch for right corners.

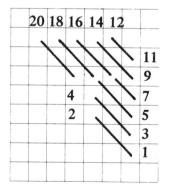

This is the slanting gobelin stitch for left corners.

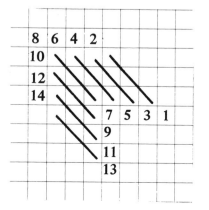

*Alternate Slanting Gobelin*

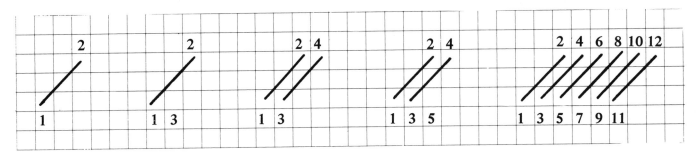

This is the alternate slanting gobelin for right corners.

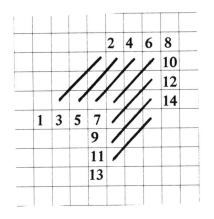

This is the alternate slanting gobelin for left corners.

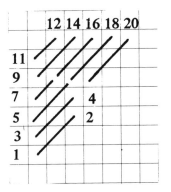

24

*St. Andrew Cross* and *St. George Cross:* These are accent stitches and should be worked last on any design that makes use of them. The St. Andrew cross looks especially good when worked alone in a small area such as the Seminole pillow described in the Seminole tribe section.

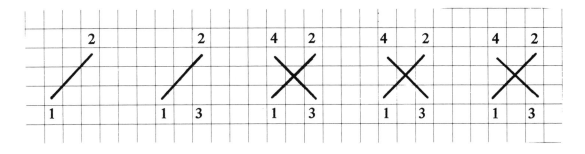

*Straight Gobelin:* This also makes a good background stitch. When worked over three mesh squares it resembles woven cloth and certain types of basketry weaving.

*Brick:* This is a series of straight gobelin stitches worked over three mesh squares and staggered to look like rows of bricks. Bricking is excellent for backgrounds or creating subtle textures. (As a left-hander, you may find it easier to work this stitch by turning your canvas a quarter turn to the left, working the stitch on its side.)

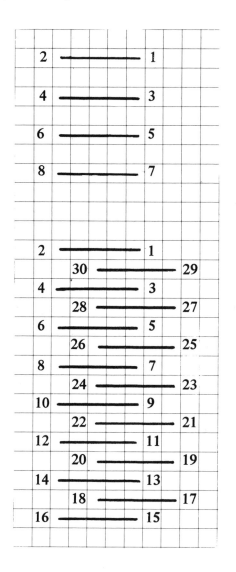

*Flame*: This, too, is a series of straight gobelin stitches worked over three mesh squares in a step pattern to form a point. It can also be worked over as many mesh squares as you desire to form either a larger or smaller point.

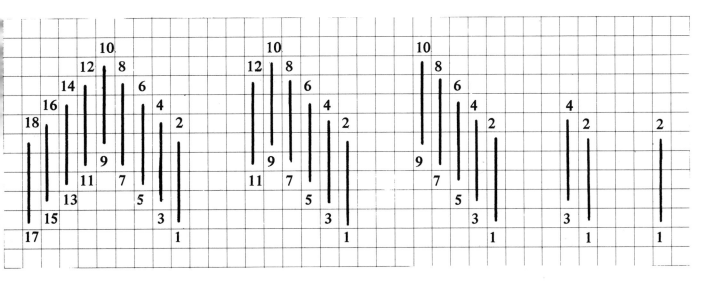

*Cherokee Diamond*: This is a stitch I created especially for the designs of the Southeastern tribes. It is worked counterclockwise, with *each* stitch overlaying the one before it. This creates a "bump" effect that works nicely with the other stitch textures. Though it may look fancy and complicated, it is easy and fun to work.

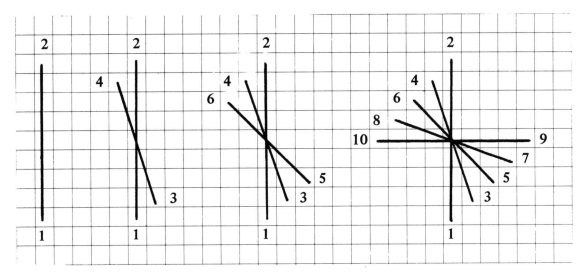

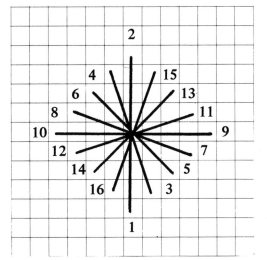

27

# 3
# KNOW YOUR CRAFT

## Yarn

Always buy too much yarn, not too little. Dye lots change frequently and many a beautiful piece has been ruined by a bit of yarn from a different dye lot with just a slightly different, but noticeable, shading. The extra yarn is always good to have, since it builds your own inventory.

Tapestry yarn does not adapt well to the stitches I have suggested in this book. The twist of tapestry yarn is very tight and does not allow the yarn to "fluff" (when no canvas threads show between the stitches). Extremely soft yarn will "pill" (fray and peel) and begin to look very worn in a short period of time. I recommend Paternayan Paterna one-hundred percent Persian wool yarn, readily available in specialty stores and department stores. I find it to be the most durable as well as attractive.

Do *not* reuse yarn that you have ripped out of your canvas. This yarn is worn and will not cover as well as new yarn.

If you are in doubt as to which shade of a color to buy when selecting yarn, always use the stronger value. It will always look more subdued when put on canvas.

The following is a selection of the colors most often used for the designs in this book. They represent the earth colors of the regions of the Southeastern Indian tribes.

| | |
|---|---|
| Red Clay | Olive Green |
| Gold | Brown |
| Leaf Green | Cobalt Blue |
| Off White | Yellow Cream |
| Pumpkin | Rust |
| Dandelion | Orange |
| Pink Cream | Lemon Yellow |
| Indigo | Maroon |

Other than the illustrations of some finished work, there are no color schemes for the designs. It is for your tastes and preferences to determine which colors you would prefer to work with and, in some cases, live with.

## Needles

For working no. 10 mesh canvas, use a no. 18 tapestry needle. To work different-sized canvas meshes, use an appropriate size needle (see table below). It is best to keep a selection of needles on hand so that you always have available the right tool for the particular project.

No. 10 mesh canvas:  No. 18 needle
No. 12 mesh canvas:  No. 18 needle
No. 14 mesh canvas:  No. 20 needle

## Accessories

Use a seam-ripper (any notions counter will have one) if you have to rip out yarn but be very careful not to rip a canvas thread.

Some other items that should be kept on hand are:

Magnifying glass—for working in small areas and looking for the proverbial "lost stitch."
Scissors.
Ruler—preferably 18 inches.
Tracing paper—for transferring designs.
Masking tape—for taping the rough edges of your canvas.
Strong light source—over the right shoulder if you are right-handed, over the left shoulder if you are left-handed.

## Canvas

The most common size canvas is no. 10 mesh. The sizes of canvas refer to the number of squares in one inch. There are three main types of canvas.

Interlock

Mono

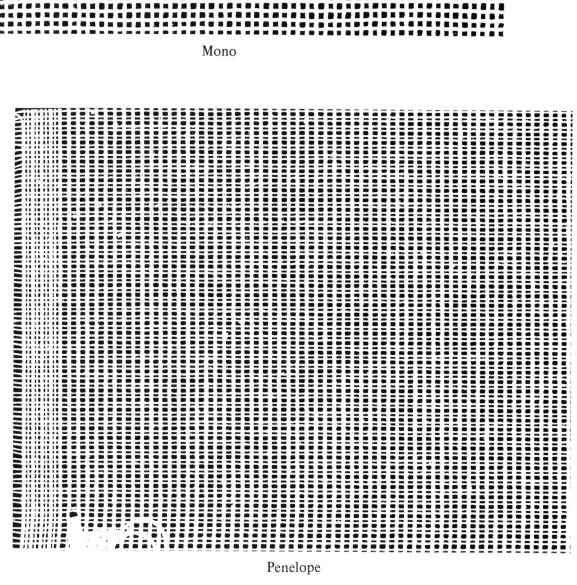

Penelope

I prefer Interlock canvas (two horizontal interwoven threads forming one crossing with two vertical interwoven threads forming one). This canvas keeps its shape and, properly worked, requires little blocking. Mono canvas (one horizontal and one vertical thread intersecting) tends to get limp as you work and distorts easily, depending upon the stitch. Penelope (two individual horizontal and two individual vertical threads at each intersection) can be difficult to work with for the beginner.

Each piece of canvas (unless cut from the center of a bolt) will have selvedge at one or two sides. Selvedge is the sides of the bolt and is the finished edge or edges when it is manufactured. The threads are "squeezed" (see illustrations) and often have a thread of color running through for easy identification. This piece will usually be your excess in working a design unless you are working the full width of a bolt of canvas, usually thirty-six, forty, or forty-two inches wide. In this case, the selvedge will be counted in your width measurement so you must be careful to allow for it when making an accurate stitch count in order to transfer a design to cover a full width.

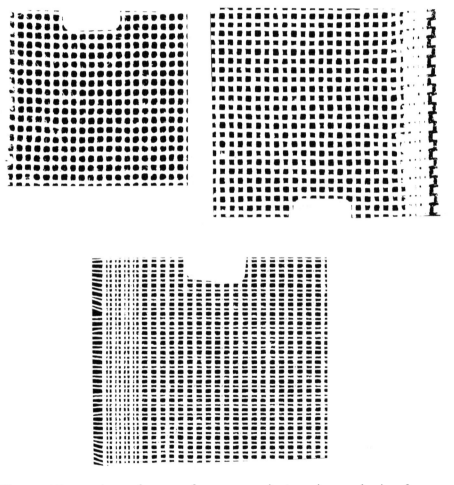

When cutting a piece of canvas for your project, or in purchasing from a store, always allow at least 1½" margin on all four sides.

Before working your piece, tape the edges of your canvas with masking tape. This prevents the yarn from snagging on the rough edges while you are working and also keeps the canvas from fraying and unraveling.

If you should have to patch a piece of canvas that is torn, always use the same size mesh and make sure that the selvedge is running in the same direction for both the patch and the piece to be patched. Many canvases will run shorter by one stitch when near the selvedge.

## Designs

Now that you have all your materials at hand, you are ready to put your design on the canvas. Speaking of materials, when putting a design on canvas, you should always use waterproof acrylic paints. They must be waterproof or else they will discolor the yarn when the finished piece is blocked.

If you hesitate to transfer the design yourself, you can take the picture or book to a photostat studio (listed in the Yellow Pages under Photo Copying). Request a *positive stat* blown up or reduced to the exact measurements you want to work with. It will be enlarged to the exact proportions and all you need to do is lay your canvas over your stat (if the lines are not dark enough, carefully go over them with a fine black pen) and, using it as a guide, slowly trace the lines onto your canvas. When you have completed this step, you can paint the design in your chosen colors, or, when you become proficient, work your colors as you go and just note your colors on your stat as a reference guide.

When tracing from your stat, or any other material for copying your design, use an ordinary lead pencil on your canvas. You will find that the light line made will wear off as you work your canvas and there will be no possibility of discoloration of the yarn when blocking.

If you want to transfer the design yourself, the procedure is the same, provided you are not enlarging or reducing the design from its original size. For instance, the designs in this book are of no. 10 mesh size and can be directly transferred to no. 10 mesh canvas. They will work up in the size that you see on the page and all you need to do is trace the design from the book onto tracing paper, place the tracing paper under your canvas, trace onto your canvas, then paint and work your piece.

If you wish to enlarge or reduce the design, you must remember one important rule: *always enlarge or reduce in proportion both horizontally and vertically.* For example, if you want to double the size of one of the designs in this book to fit a large pillow, draw your vertical lines as two inch for every one inch. Do the same horizontally. This will give you a box that is exactly twice the size as the original. Now, line off the box into squares and match the squares in the original to the squares in your enlargement. Any line you find in one of the small squares, place in the corresponding large square. Follow this pattern and you will soon transfer the design, square by square. If your design is too large, simply reverse the process. Since the designs in this book are geometrics, you will find them easy to transfer using this small-square-to-large-square method.

When you have accomplished this step, take your transferred design, place it under your canvas, and follow the same steps described above.

## Working Your Canvas

When you begin working your design, start in the center and work the detail first. The background should be the last section worked, since it is the largest and will often be easier.

Remember that most stitches begin at the bottom of the stitch and go into the canvas at the top of the stitch. This is especially true of the straight stitches.

When working the design, *put the needle through one canvas square at a time.* This method may take a little more time, but will save you a great deal of work later, since your piece will require less blocking when finished. This method allows you to work with regular and even tension (see illustration).

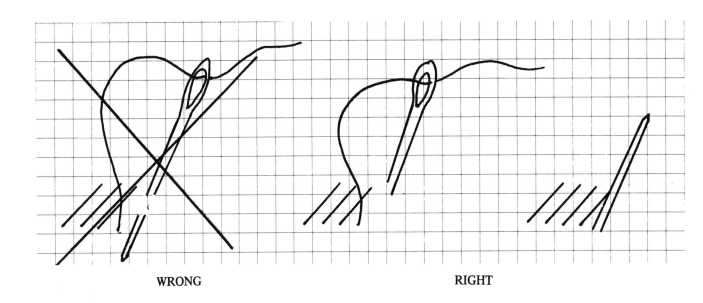

WRONG                                    RIGHT

Keep an eye on your tension. A bit of practice will help keep your stitches of even and slightly loose tension. Uneven tension will warp the canvas in many directions and create a problem that can only be solved by a tedious and often expensive blocking job.

*Roll your canvas — never fold it.* Folds cause ridges that may not come out in blocking.

Roll the canvas in your hand as you are working. This will also keep it from warping, avoid folds, and help keep your stitches even.

Avoid splitting your yarn with the needle when you go into a mesh square that has already been used in another stitch. It will result in a weakened piece of yarn that will soon fray or break.

When your yarn begins to tighten its natural twist, simply give it a half turn of your wrist in the opposite direction, or let the needle and yarn hang free and they will untwist themselves.

Never knot your yarn. This causes bumps on your canvas that cannot be blocked out. The ends of the strands of yarn should be woven into the work on the back of the canvas so that no end hangs loose and can be pulled from the front of the canvas (see illustration).

34

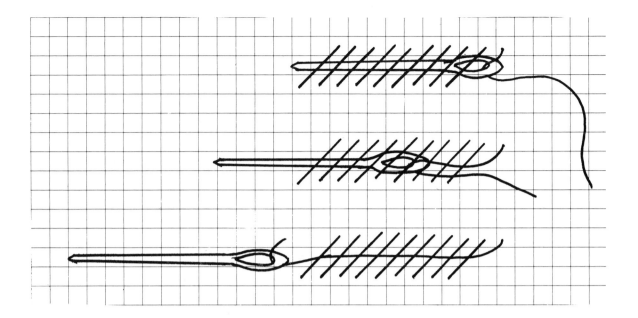

Occasionally a compensating or partial stitch will be needed. This will occur when there are not enough canvas threads for a full stitch. The compensating stitch is that part of a stitch which will cover the remaining area. These stitches can be put in while you are working or as finishing touches. As you work your design, you will decide which way suits you best (see illustration).

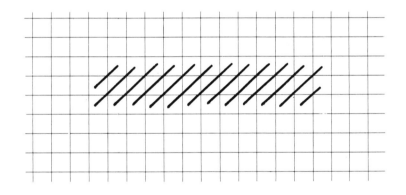

When you finish a work session with your piece, secure the needle on the excess canvas around your design. (Remember the extra 1½" margin you have on your canvas.) To keep from warping any portion of your design, put the needle as close to the edge of your margin allowance as is possible.

## The Finishing Touches

All of the stitches in this book are long stitches, usually covering more than one mesh square. They should be worked with light, even tension and slightly loose. The finished work may only require a minimum of blocking.

If this is the case, a little steam can bring out the "fluff" in the yarn. Place your finished piece, face down, on a bath towel. Hold the steam iron just over the back of the piece and release the steam. It will penetrate the yarn. Do not iron the needlework under any circumstances, since you will flatten and distort the yarn.

When your piece requires blocking, place it, *face up,* on a wooden board. Using aluminum or rust-proof tacks, fasten your piece to the board in a square or rectangle. To wet your piece, use some form of sprayer that emits a mist rather than a heavy spray. Do not soak the piece to the point of having excess water. As you spray, stretch the piece until you are satisfied that you have removed the wrinkles and straightened all edges. Leave it to dry in a horizontal position. When it is completely dry, remove it and check your results. In some cases, you may have to repeat the process a few times, but, if you follow the tips I have given you previously, your work should be ready to be finished and mounted.

I have not included any instructions on finishing, mounting, or framing for there are many different methods to follow. If you do not want to take your work to a professional, remember, if you do it yourself, use a little common sense and a lot of patience and you can do it. A little sewing knowledge will help you finish your work.

Finally, when you take on a project from start to finish, it is important that you develop good working habits. The tips and techniques I have described on the previous pages are only as good as your willingness to use them and your concentration on your work.

From here, you are on your own. Good luck.

# 4
# THE DESIGNS

All of the designs on the following pages have been adapted from or inspired by the authentic work of the Southeastern tribes. The captions indicate the source of the original work and, wherever possible, any significance the original may have had to the culture of the tribe. Many of these designs have been taken directly from a piece of pottery, basketry, textile, or some other Indian artifact with the only changes being the "adaptation" to make it suitable for needlework. Finally, there are some pre-Columbus (prior to the discovery of America in 1492) designs adapted from fragments of artifacts found at various archaeological excavations. The identification to a tribe is based upon the general locale of the discovery and some "educated guesses" by the authorities making the find. It also required some "educated guesses" on my part, along with some archaeological drawings in some cases, to complete the design.

*Design* 1.    Cane mat design. Traditional motif called "Broken Heart."

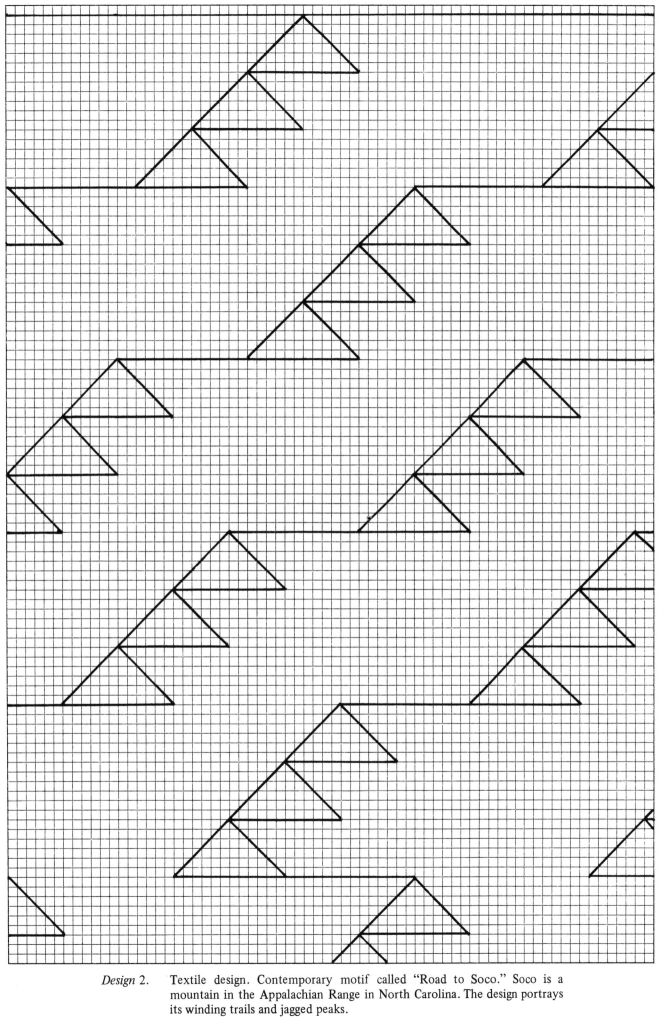

*Design* 2.     Textile design. Contemporary motif called "Road to Soco." Soco is a mountain in the Appalachian Range in North Carolina. The design portrays its winding trails and jagged peaks.

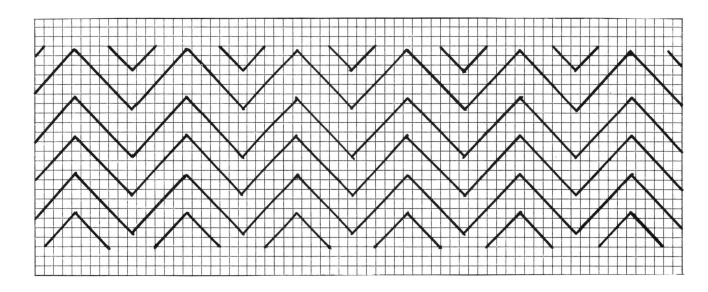

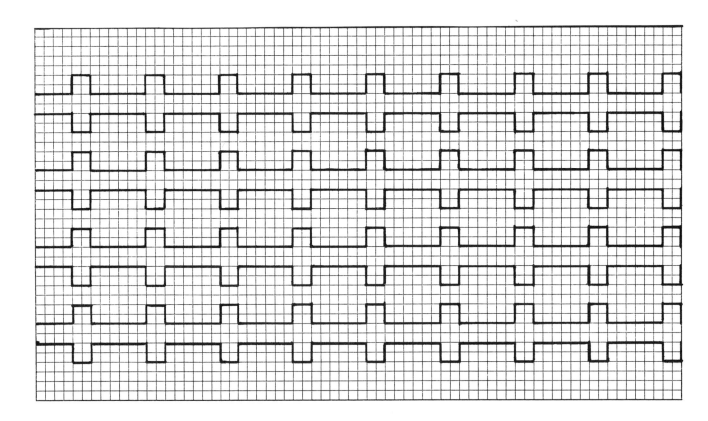

*Design* 3.　　(Top) Basket design. Continuous zigzag motif.
　　　　　　　　(Bottom) Basket design. Key motif.

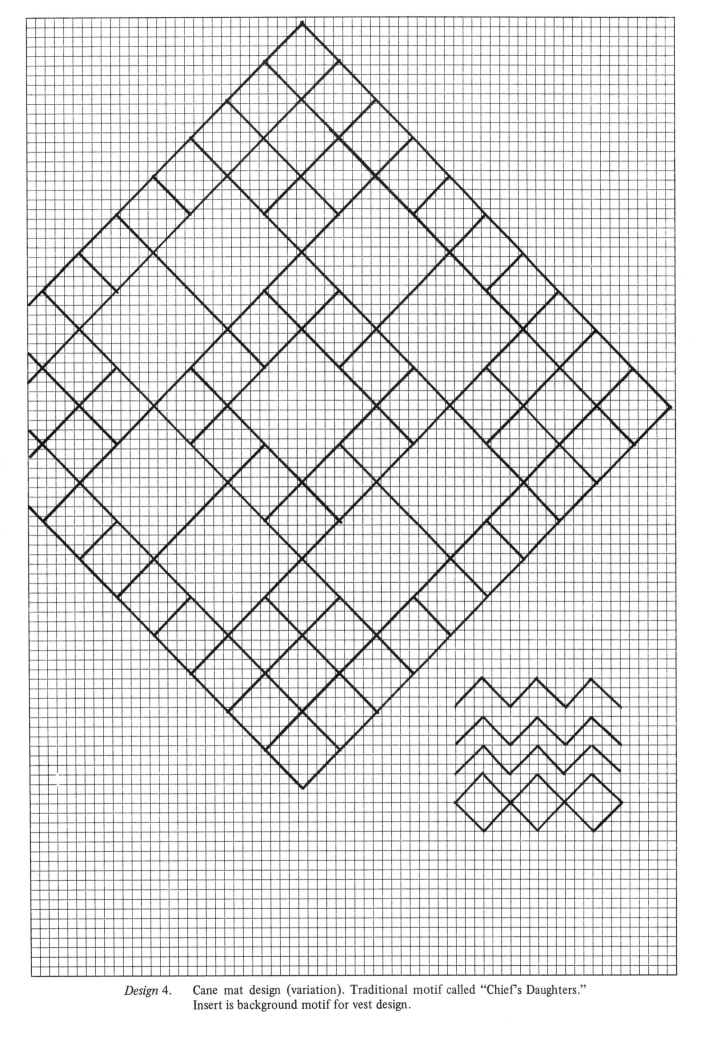

*Design* 4. Cane mat design (variation). Traditional motif called "Chief's Daughters."
Insert is background motif for vest design.

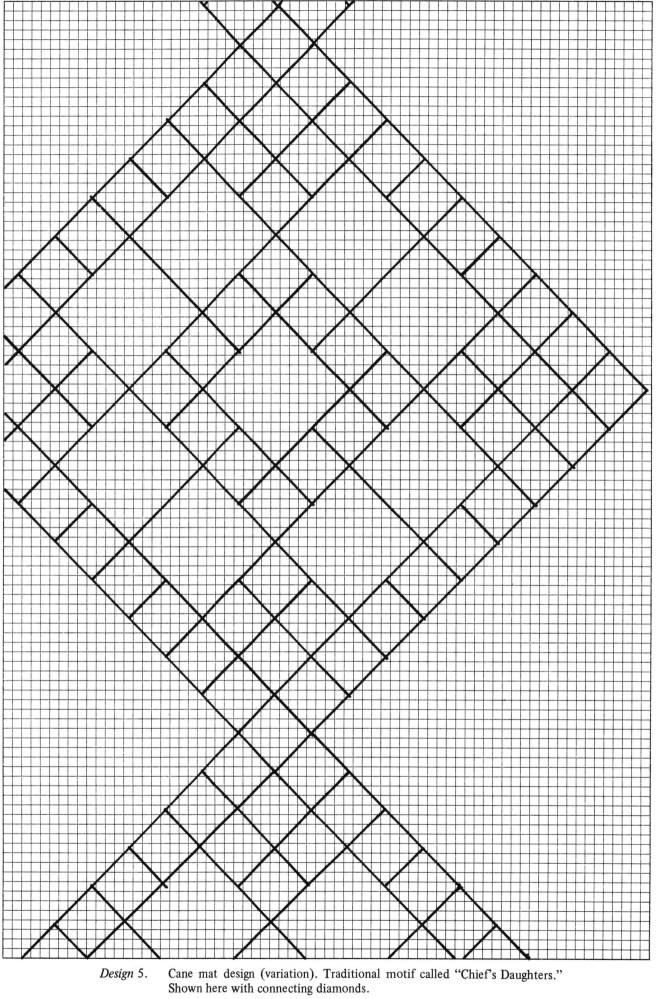

*Design* 5.  Cane mat design (variation). Traditional motif called "Chief's Daughters."
Shown here with connecting diamonds.

*Design* 6.    Cane mat design. Traditional motif called "Man in the Coffin."

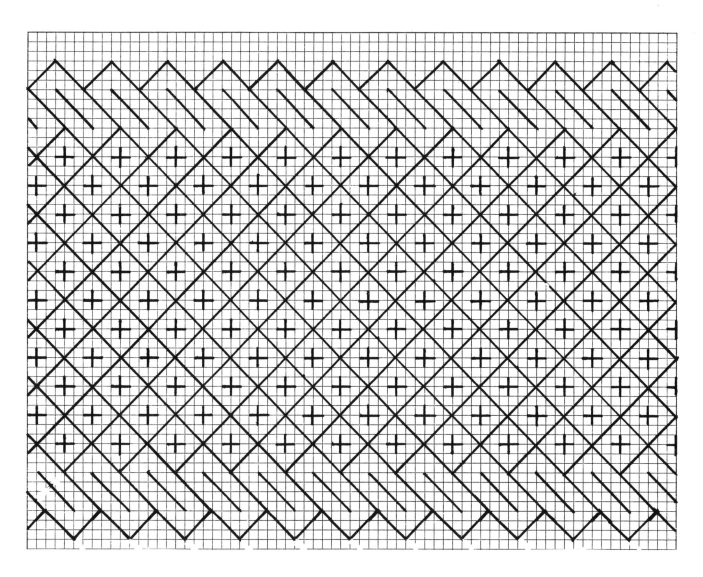

Design 7. (Top) Pottery design. Combination of "Man in the Coffin" on the outside, "Chiefs Daughters" on the inside.
(Bottom) Basket design.

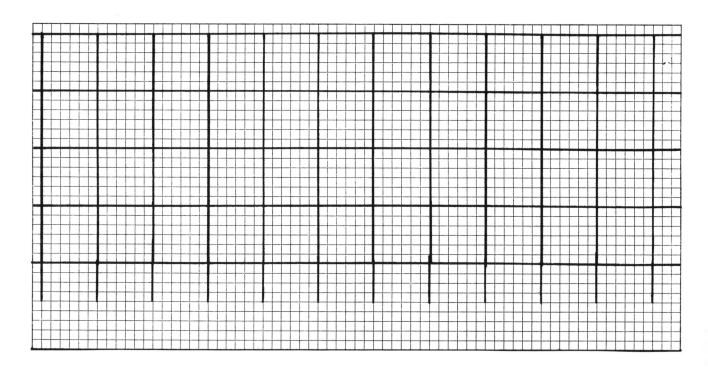

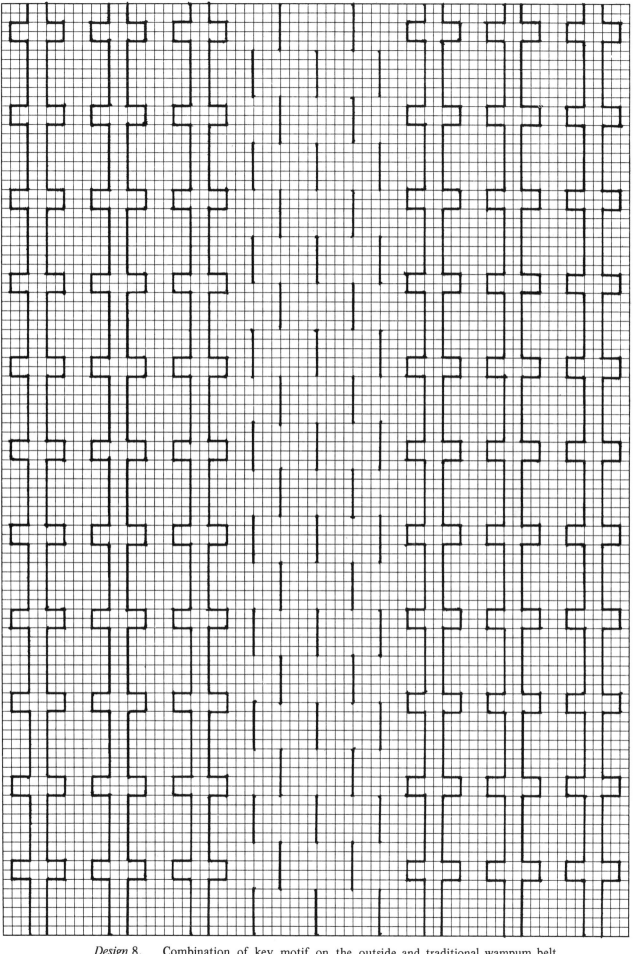

*Design* 8.    Combination  of  key  motif  on  the  outside  and  traditional  wampum  belt
motif on the inside.

*Design* 9.    Basketry designs.

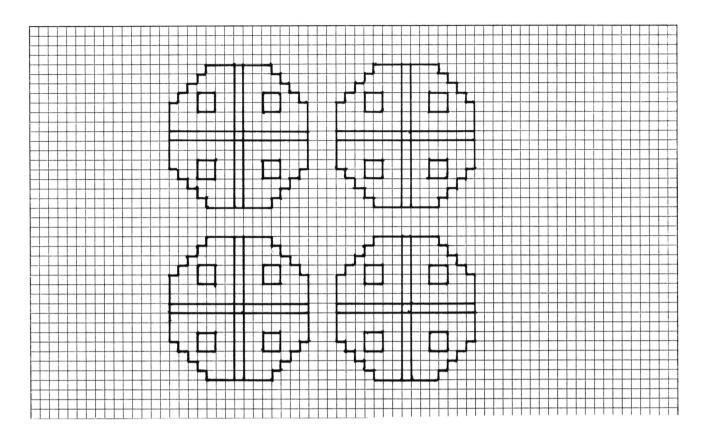

*Design* 10.   (Top) Wooden paddle design used to stamp motif on pottery.
(Bottom) Basketry design.

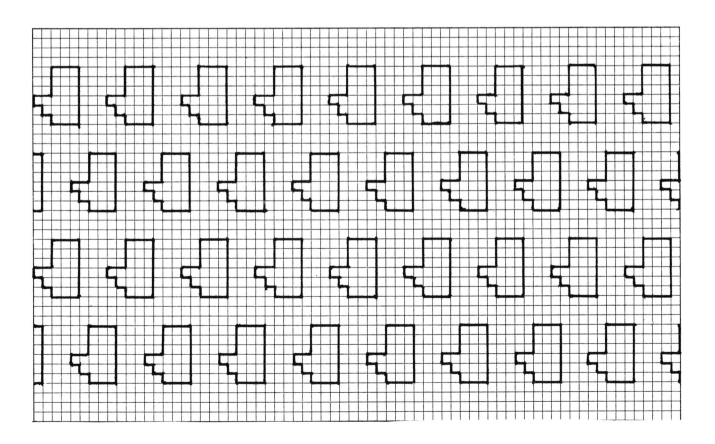

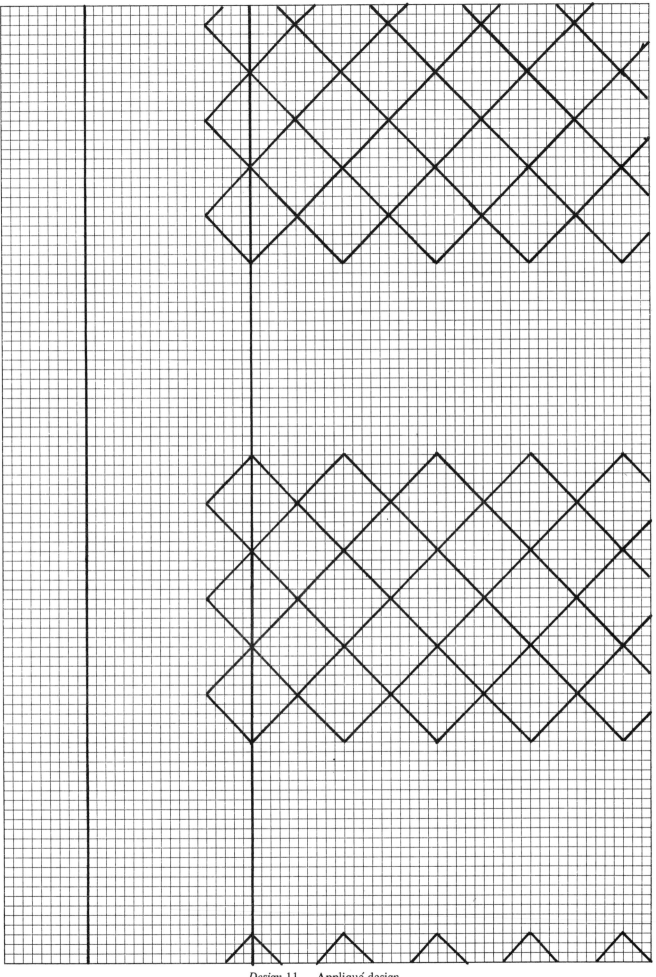

*Design* 11.   Appliqué design.

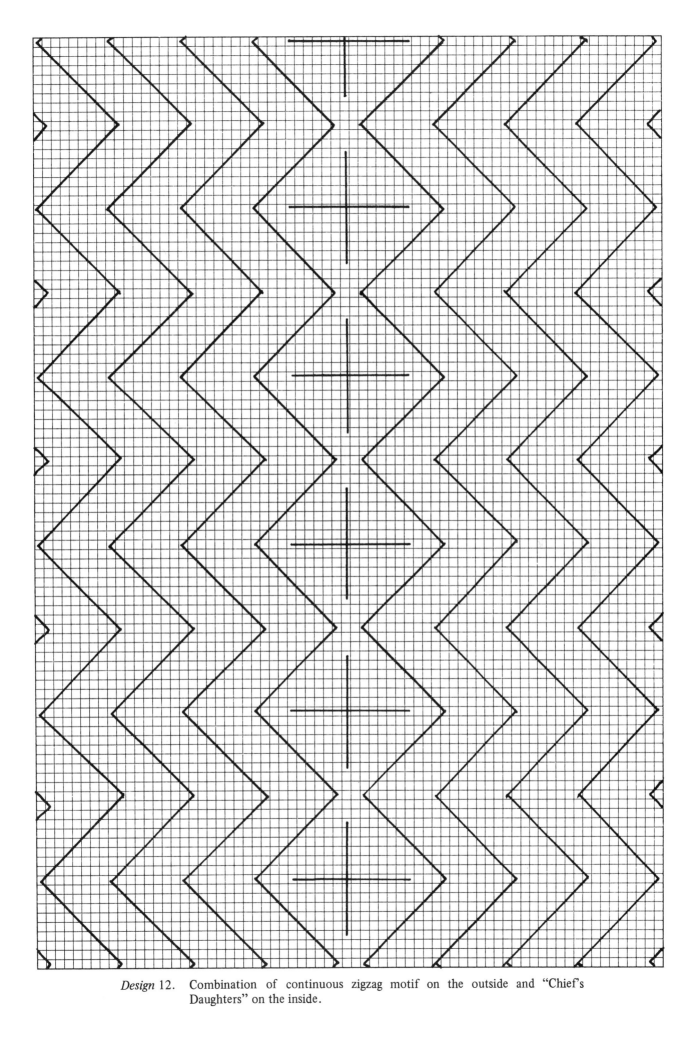

*Design* 12. Combination of continuous zigzag motif on the outside and "Chief's Daughters" on the inside.

*Design* 13. Pottery design. Combination of zigzag and "Chief's Daughters" motifs interwoven.

*Design* 14.    Pottery designs with variations on border motifs.

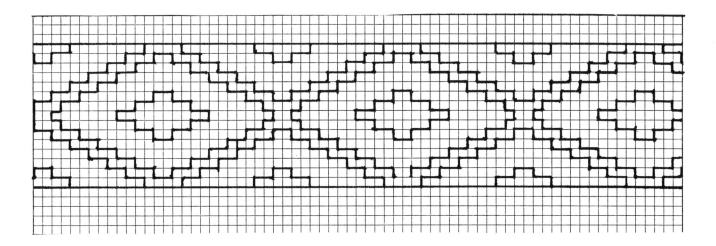

Design 15. (Top) Shoulder sash design.
(Bottom) Traditional wampum (money) belt design.

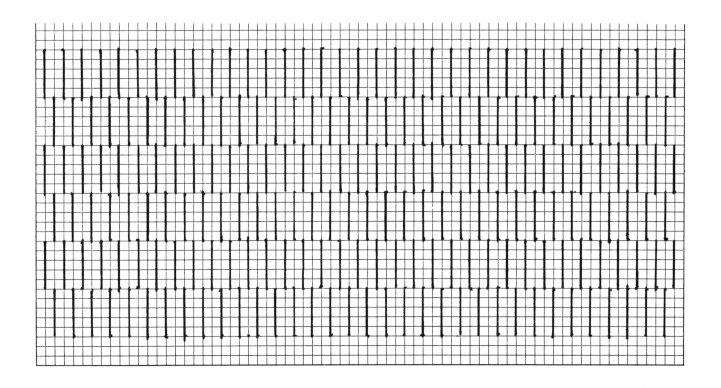

*Design* 16.    Pottery designs. Combination of zigzag and "Chief's Daughters" motifs interwoven.

*Design* 17.   Cane mat design. Combination of zigzag and "Chief's Daughters" motifs.

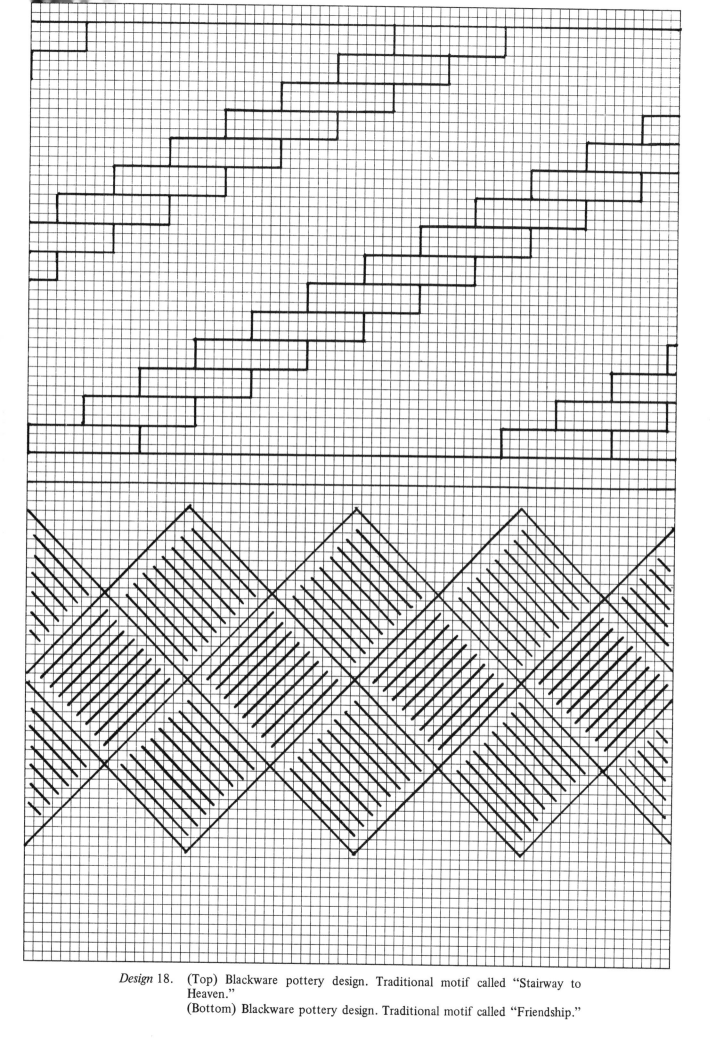

*Design* 18.   (Top) Blackware pottery design. Traditional motif called "Stairway to Heaven."

(Bottom) Blackware pottery design. Traditional motif called "Friendship."

*Design* 19.    Basketry designs. Variations on traditional "Man in the Coffin" motif.

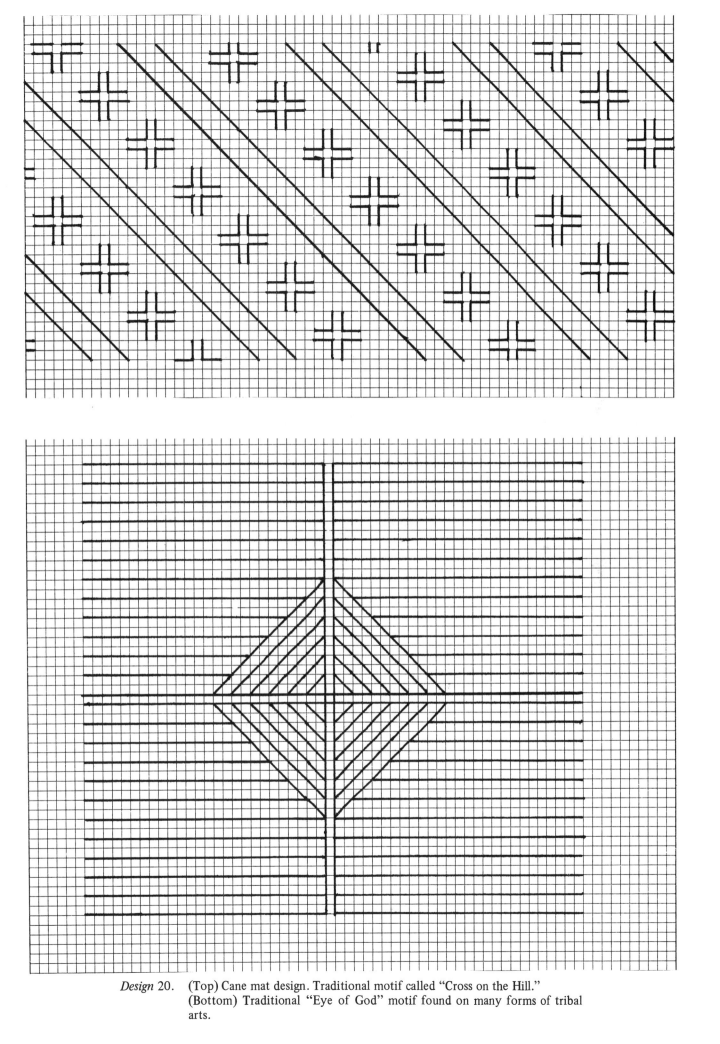

*Design* 20. (Top) Cane mat design. Traditional motif called "Cross on the Hill."
(Bottom) Traditional "Eye of God" motif found on many forms of tribal
arts.

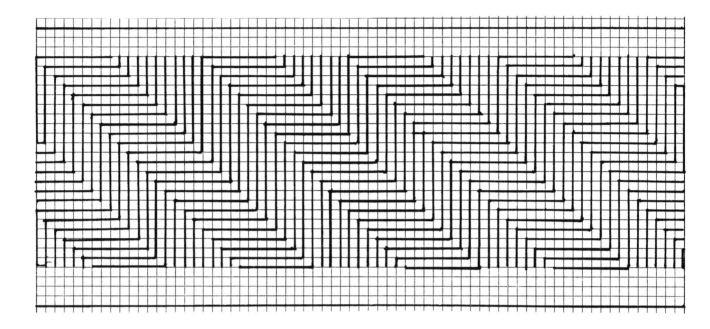

Design 21. (Top) Basketry constructon technique called "twilling."
(Bottom) Belt and textile design.

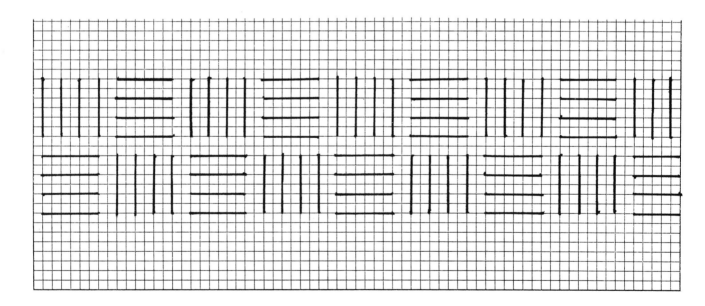

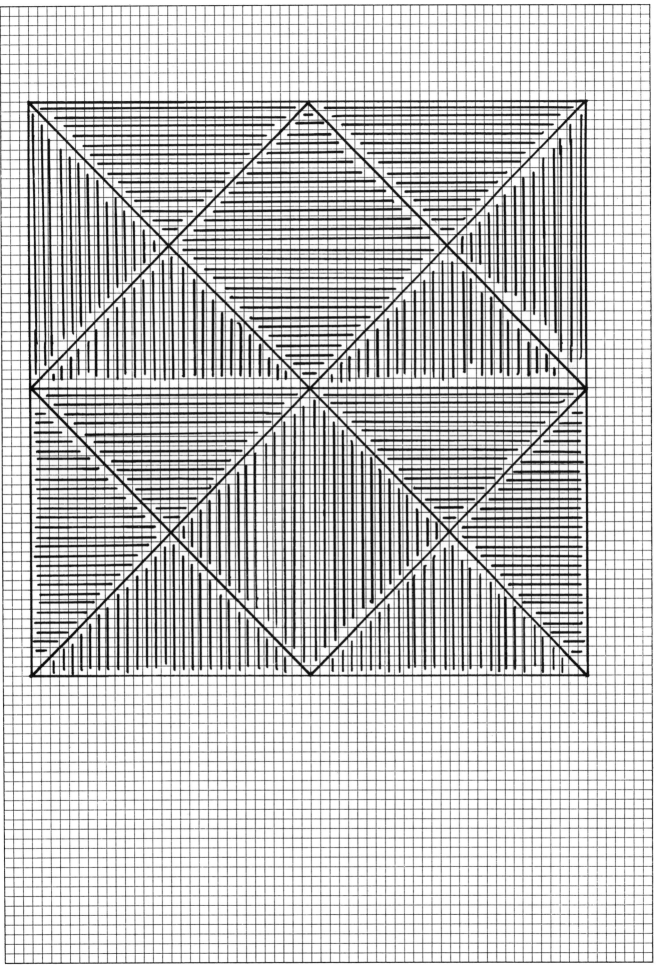

*Design* 22.   Wooden paddle design used to stamp motifs on pottery.

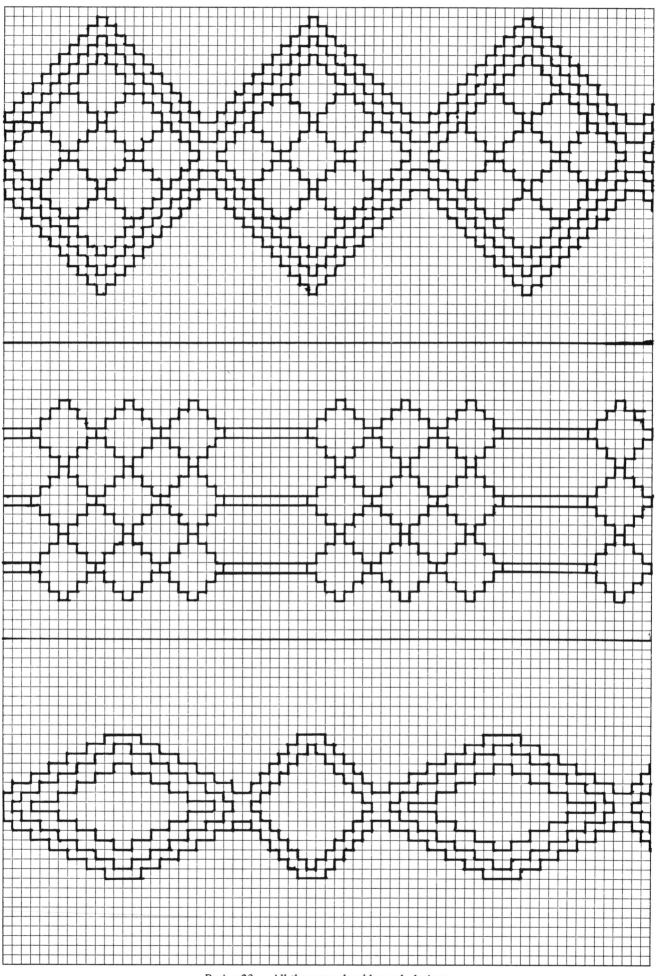

*Design* 23.   All three are shoulder sash designs.

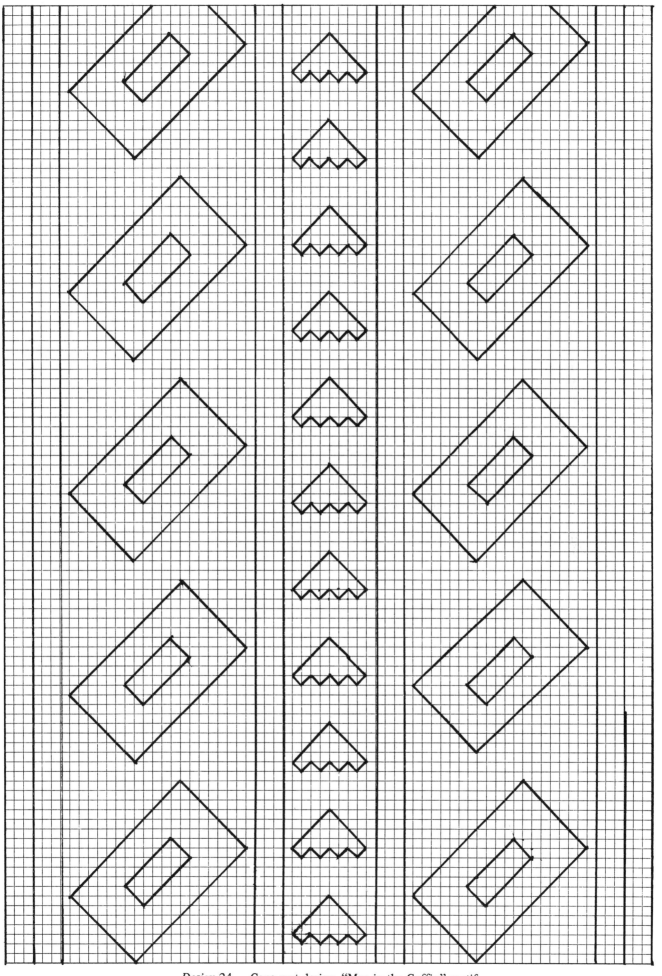

*Design* 24.    Cane mat design. "Man in the Coffin" motif.

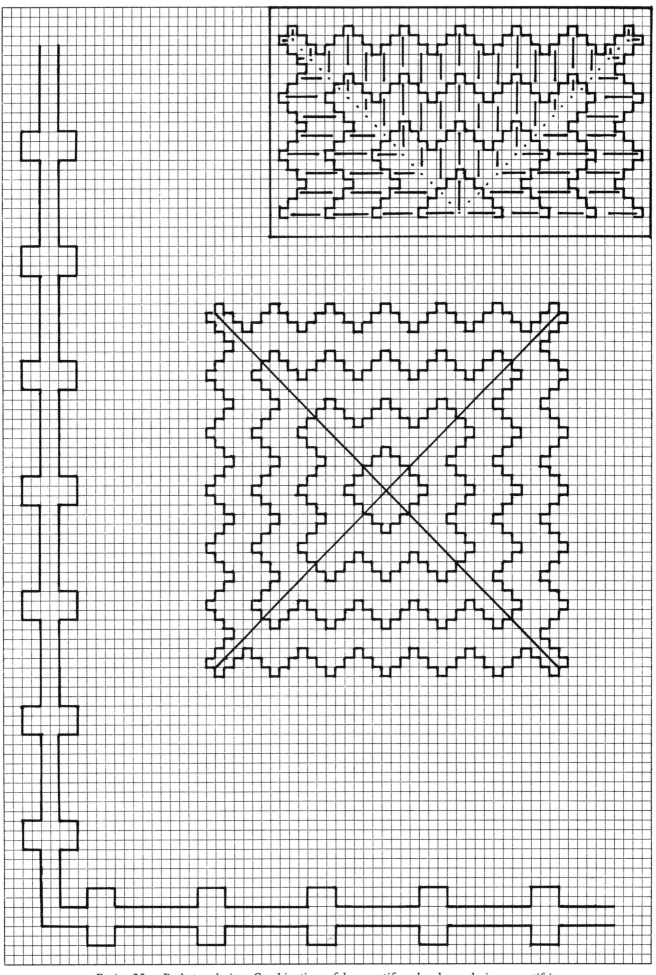

*Design* 25.   Basketry design. Combination of key motif on border and zigzag motif in center. Insert is detail of center motif.

*Design* 26.   Wooden paddle design used to stamp motifs on pottery.

*Design* 27.   Wooden paddle design used to stamp motifs on pottery.

Cherokee chair. (See Design 26.)

Cherokee vest. (See Designs 4 and 5.)

Cherokee vest. (See Designs 4 and 5.)

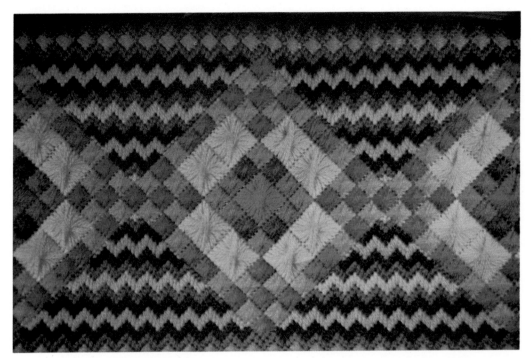

Cherokee vest. (See Designs 4 and 5.)

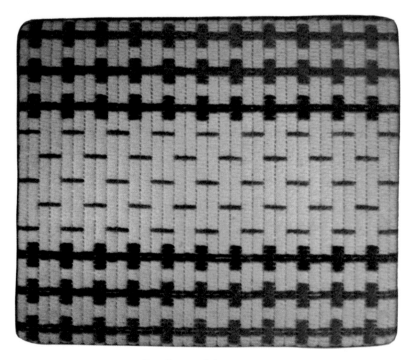

Cherokee wall hanging. (See Design 8.)

Cherokee wall hanging. (See Design 13.)

Cherokee wall hanging. (See Designs 3, 27, 28.)

Cherokee glasses case (side 1).
(See Design 22.)

Cherokee glasses case (side 2). (See Design 10.)

Cherokee picture, framed. (See Design 22.)

Creek bookmark (left). (See Design 41.)

Creek belt. (See Design 41.)

Choctaw chair. (See Design 70.)

Choctaw sandals. (See Design 66.)

Choctaw handbag. (See Design 78.)

Seminole pillow. (See Designs 82, 83, and 84.)

Seminole chair. (See Design 96.)

Seminole chair. The design is worked directly on the back and seat of the chair, using the cane as you would a piece of canvas. (See Design 96.)

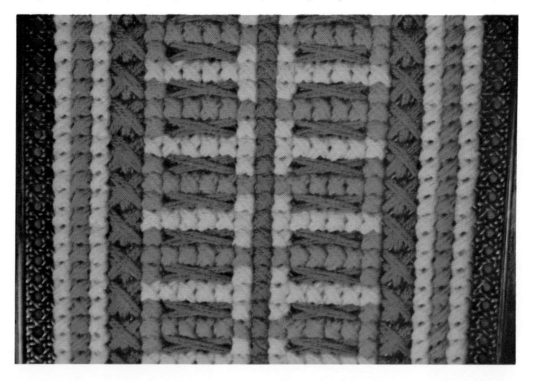

Cherokee blackware bowl—paddle-stamped decoration. *Courtesy Museum of the American Indian, Heye Foundation.*

Creek beaded shoulder bag. *Courtesy Museum of the American Indian, Heye Foundation.*

Choctaw woven cane basket, loop handle. *Courtesy Museum of the American Indian, Heye Foundation.*

Seminole male and female dolls in costume. *Courtesy Museum of the American Indian, Heye Foundation.*

*Design* 28.   Pottery design. Incised or cut into pottery instead of stamped.

*Design* 29. Shell gorget (necklace) design. Found in prehistoric site in Eastern Tennessee and believed to be of Cherokee origin.

*Design* 30.    Same design as the preceding page. This is a further adaptation of the design for needlepoint.

**Creek**

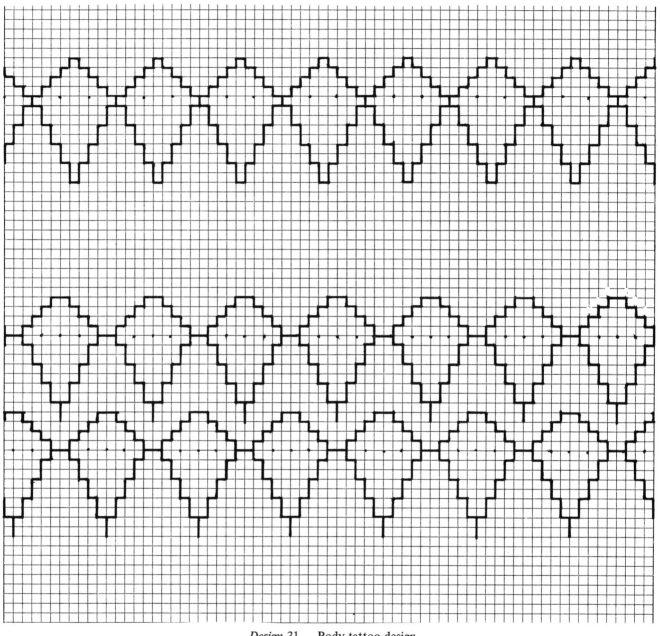

*Design* 31.   Body tattoo design.

*Design* 32. (Top and center) Shoulder bag designs.
(Bottom) Body tattoo design.

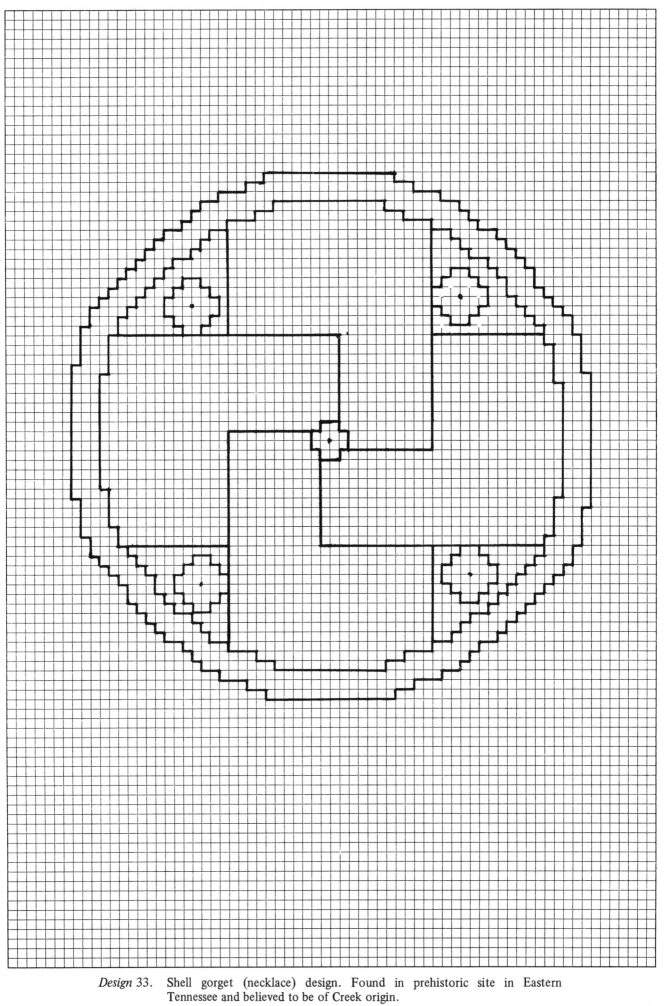

*Design* 33.   Shell gorget (necklace) design. Found in prehistoric site in Eastern Tennessee and believed to be of Creek origin.

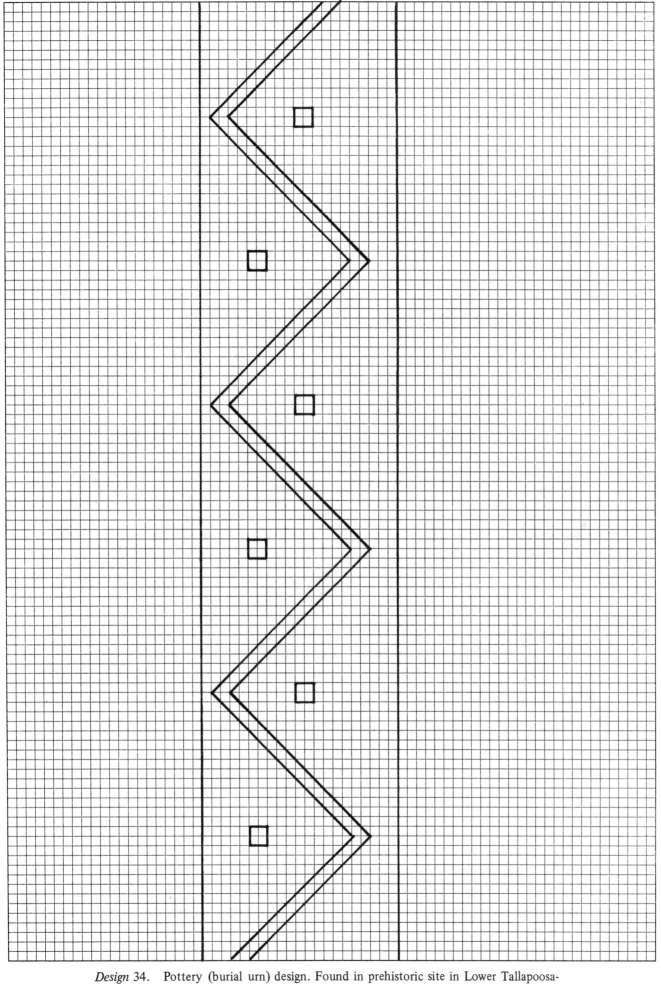

*Design* 34. Pottery (burial urn) design. Found in prehistoric site in Lower Tallapoosa-Upper Alabama River area and dated to Mississippi Cultural period (A.D. 1300-1600) and believed to be of Creek origin.

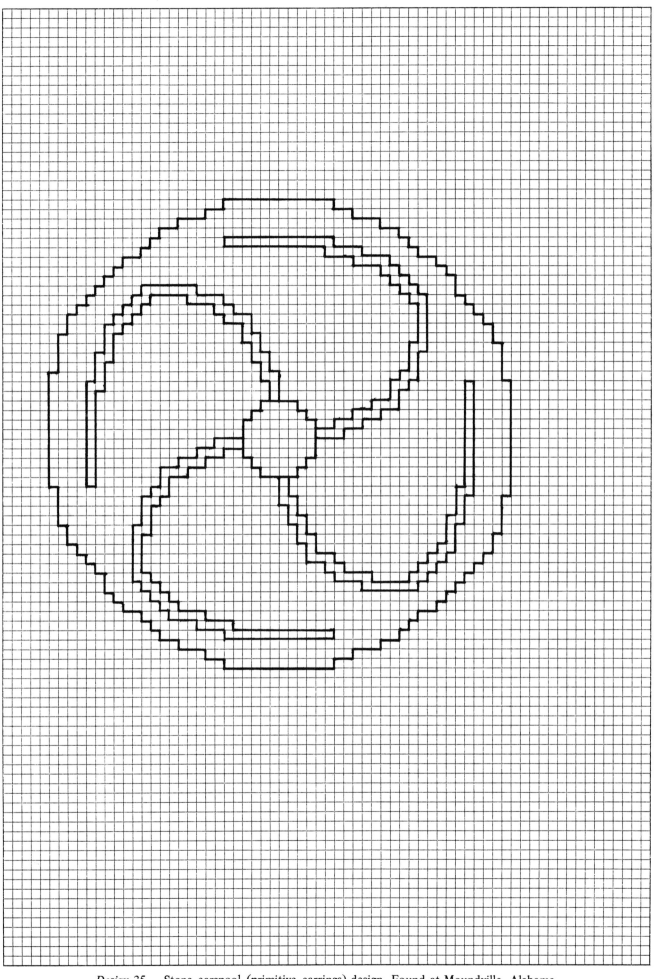

*Design* 35.    Stone earspool (primitive earrings) design. Found at Moundville, Alabama, excavation and believed to be of Creek origin.

*Design* 36.   Beaded shoulder bag designs. Traditional arrowhead motifs.

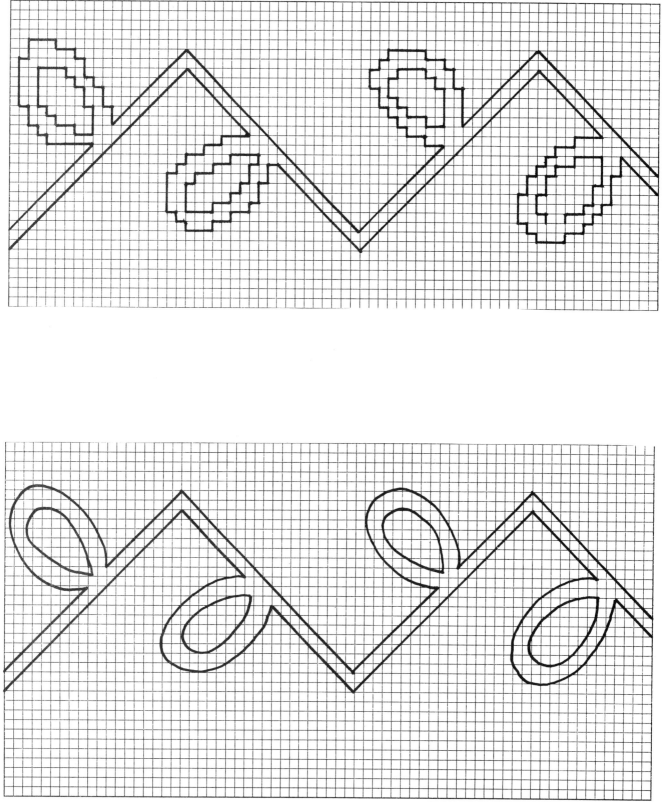

*Design* 37.   Shoulder bag designs.

*Design* 38.    Shoulder bag design.

*Design* 39.   Shoulder bag design.

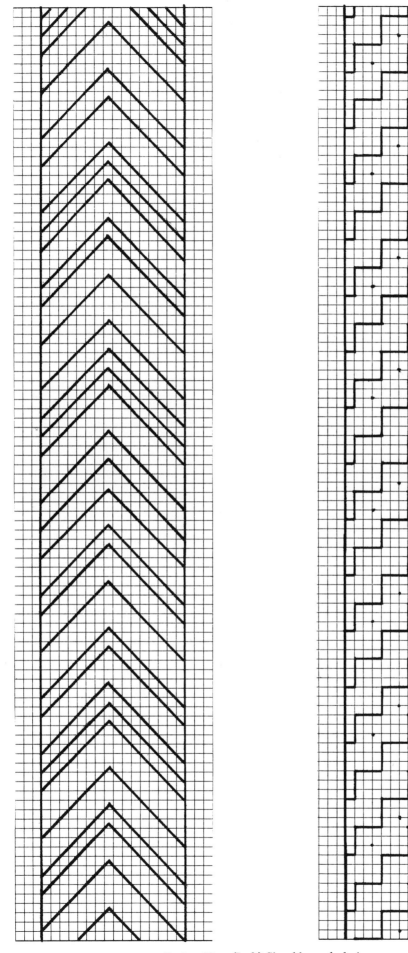

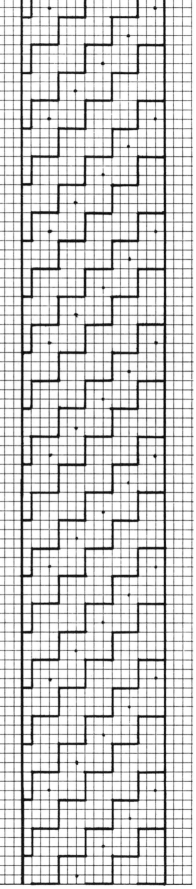

*Design* 40.   (Left) Shoulder sash design.
(Right) Body tattoo design.

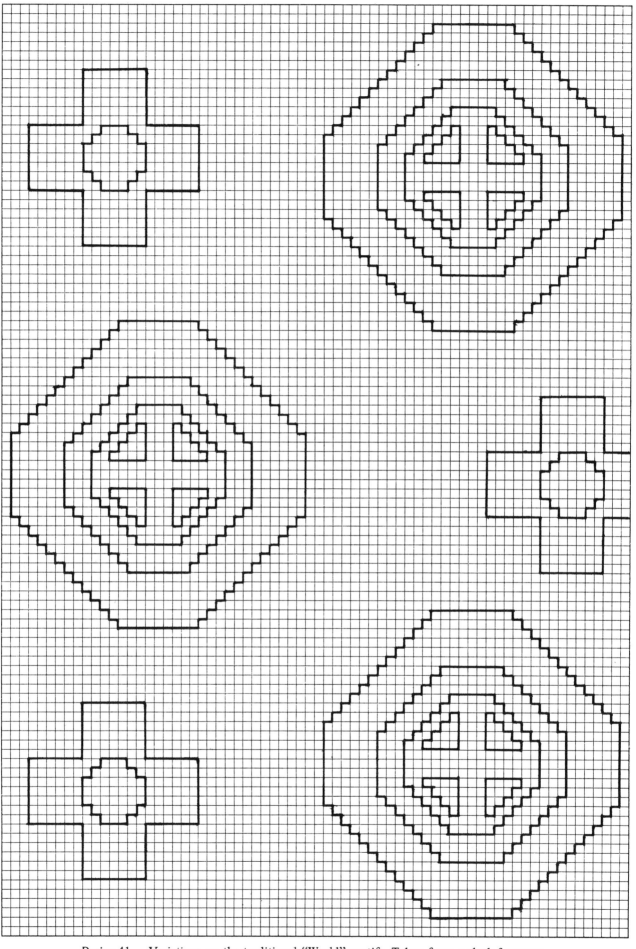

*Design* 41.   Variations on the traditional "World" motifs. Taken from a cloth fragment found at the Etowah site excavation in Georgia and believed to be of Creek origin.

*Design* 42.   (Top) Beaded sash design.
(Bottom) Facial tattoo design.

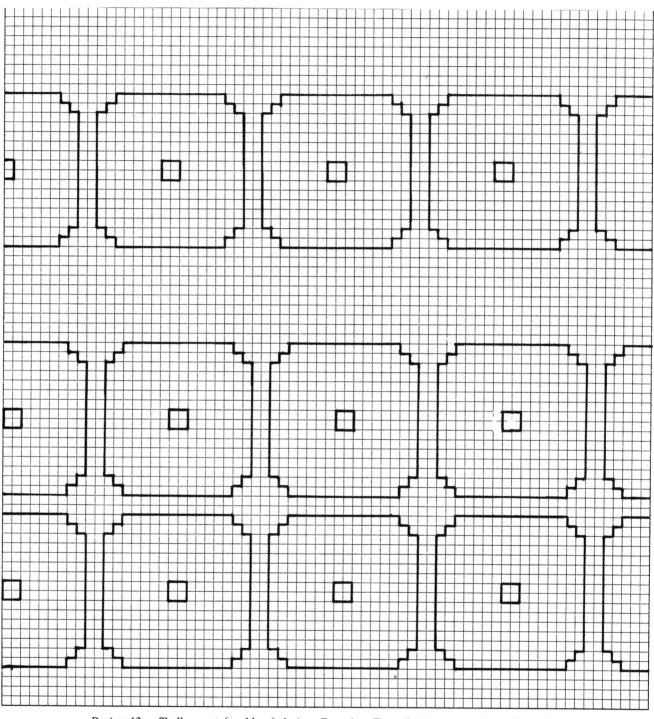

*Design* 43.   Shell gorget (necklace) design. Found at Etowah site excavation in Georgia and believed to be of Creek origin.

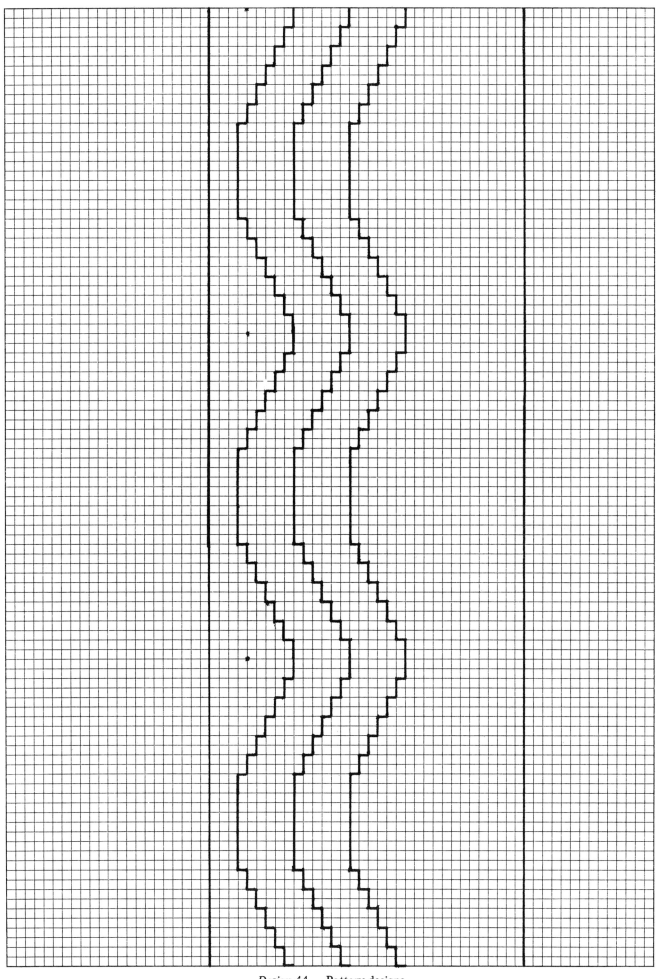

*Design* 44.    Pottery designs.

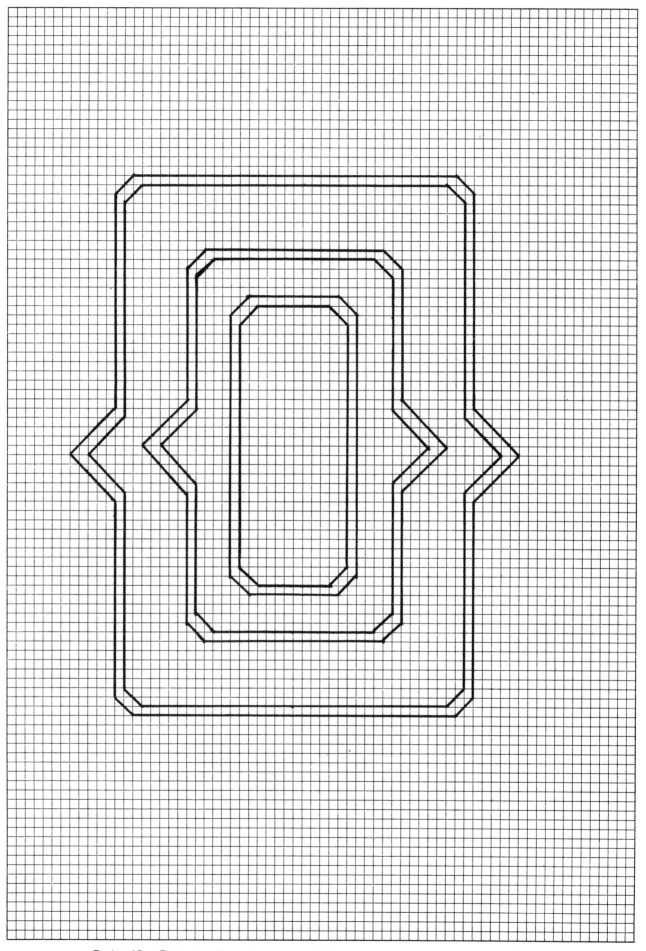

*Design* 45.    Pottery and textile design. Traditional motif called "Open Eye," found at excavation, Moundville, Alabama and believed to be of Creek origin.

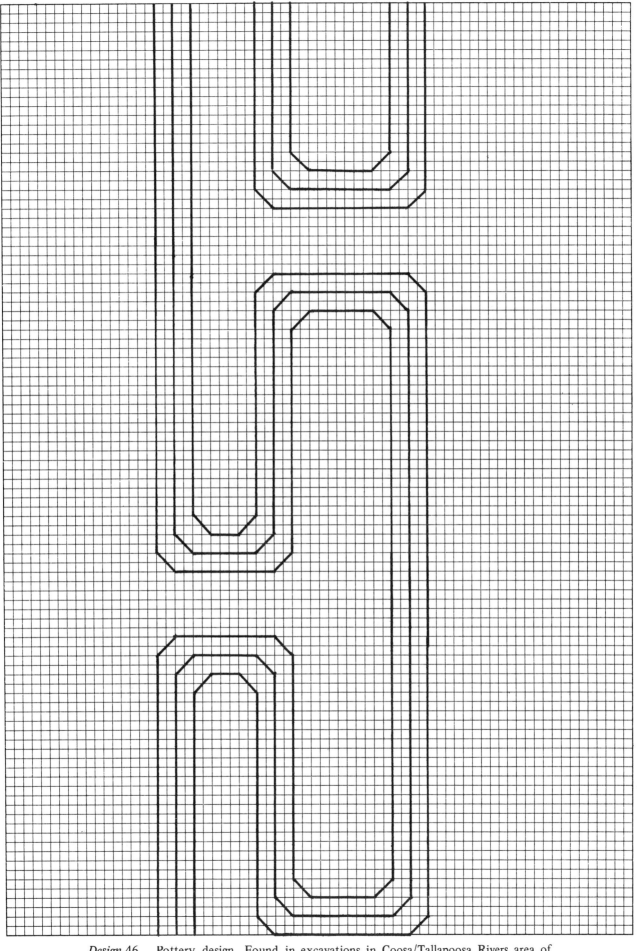

*Design* 46.   Pottery design. Found in excavations in Coosa/Tallapoosa Rivers area of
Alabama and believed to be of Creek origin.

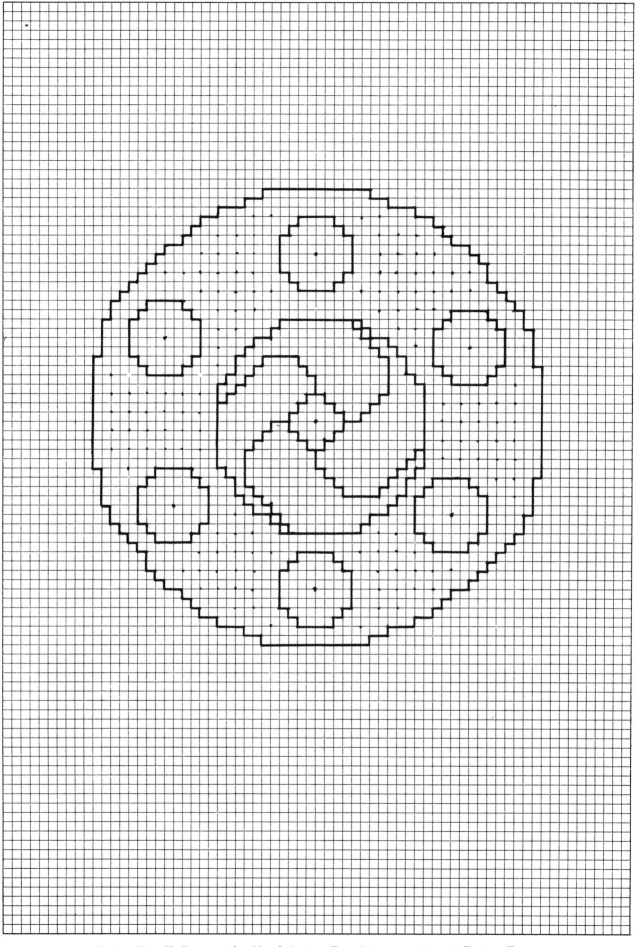

*Design* 47.    Shell gorget (necklace) design. Found in excavations in Eastern Tennessee and believed to be of Creek origin.

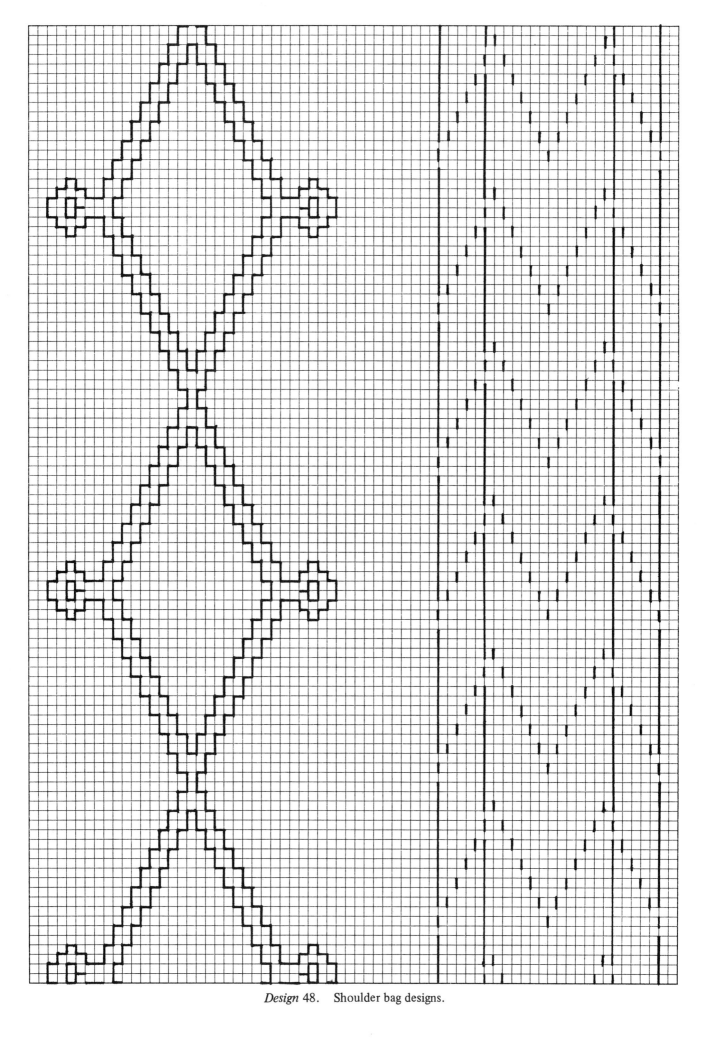

*Design* 48. Shoulder bag designs.

*Design* 49. Pottery (bowl) design. Traditional "Hand and Eye" motif found engraved on a bowl fragment at excavation, Moundville, Alabama and believed to be of Creek origin.

*Design* 50.   Pottery design. Traditional "Hand and World" motif found painted on a glass bottle around Nashville, Tennessee, and believed to be of Creek origin.

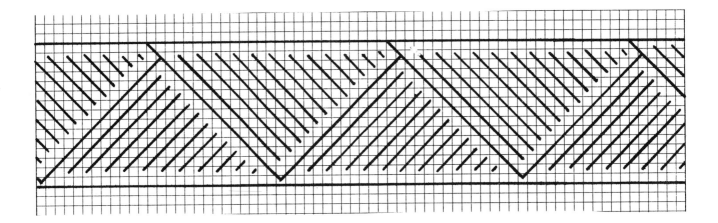

Design 51. (Top) Pottery design.
(Bottom) Shoulder bag design.

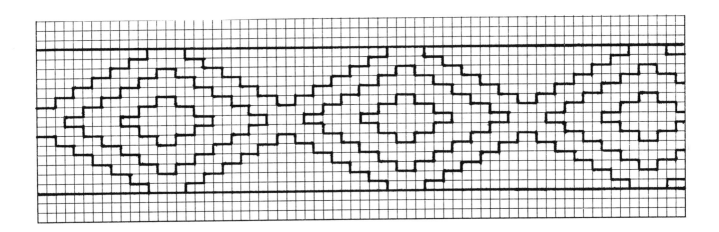

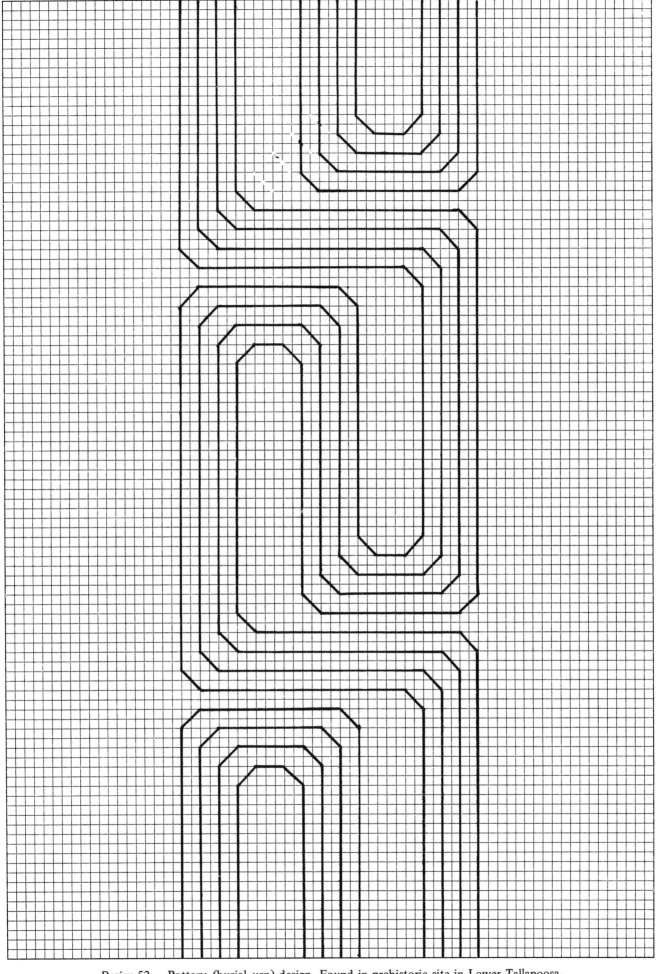

*Design* 52.   Pottery (burial urn) design. Found in prehistoric site in Lower Tallapoosa-
Upper Alabama River area and dated to be Mississippi Cultural period (A.D.
1300-1600) and believed to be of Creek origin.

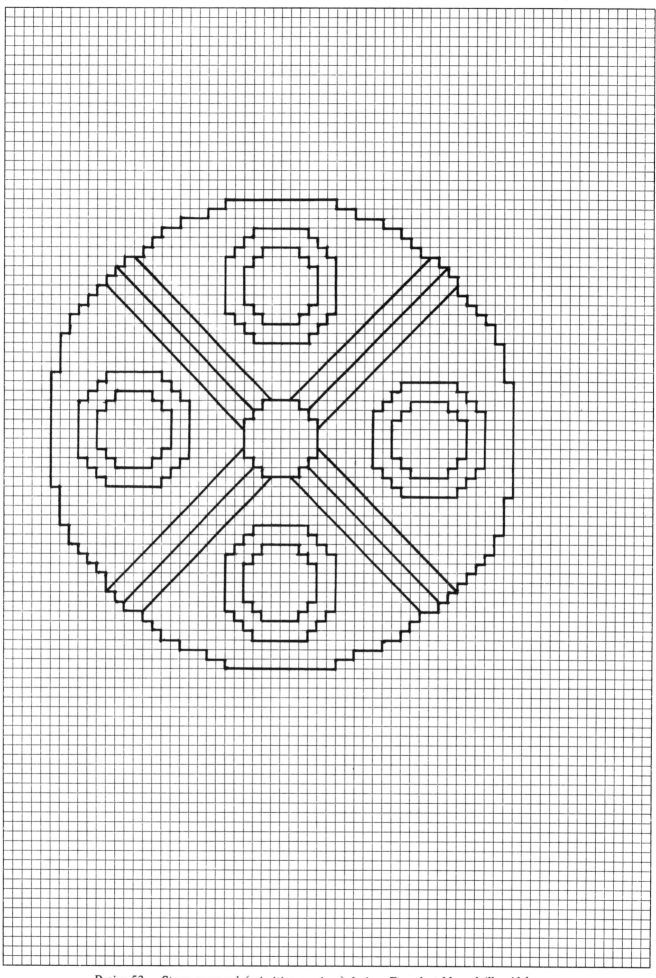

*Design* 53.  Stone earspool (primitive earrings) design. Found at Moundville, Alabama, excavation and believed to be of Creek origin.

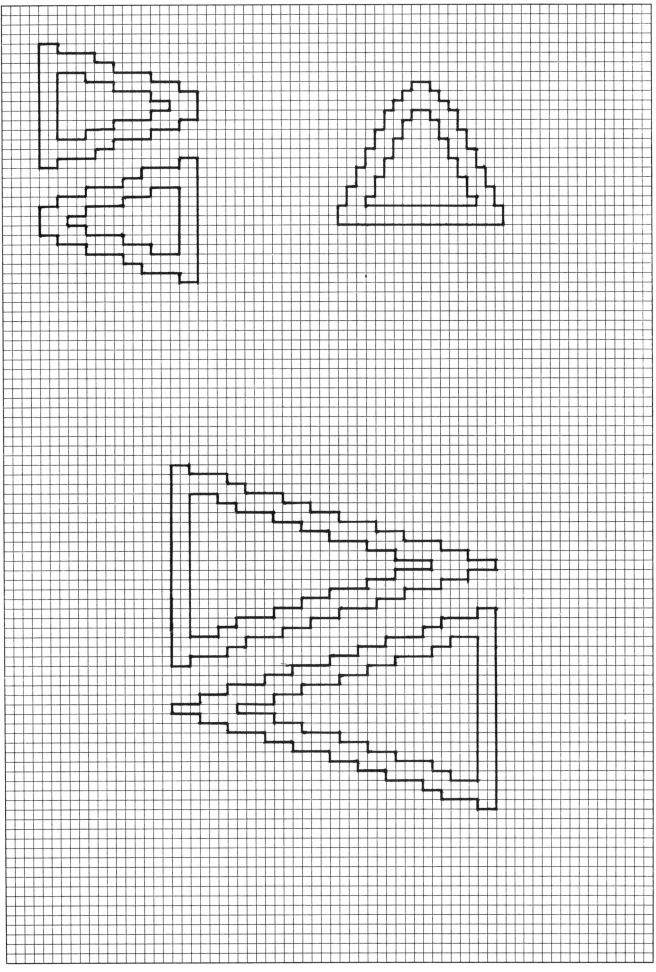

*Design* 54. Beaded shoulder bag design. Traditional arrowhead motifs.

*Design* 55.    Shell gorget (necklace) design. Similar gorgets have been found in Northern
Alabama and like this gorget, believed to be of Creek origin.

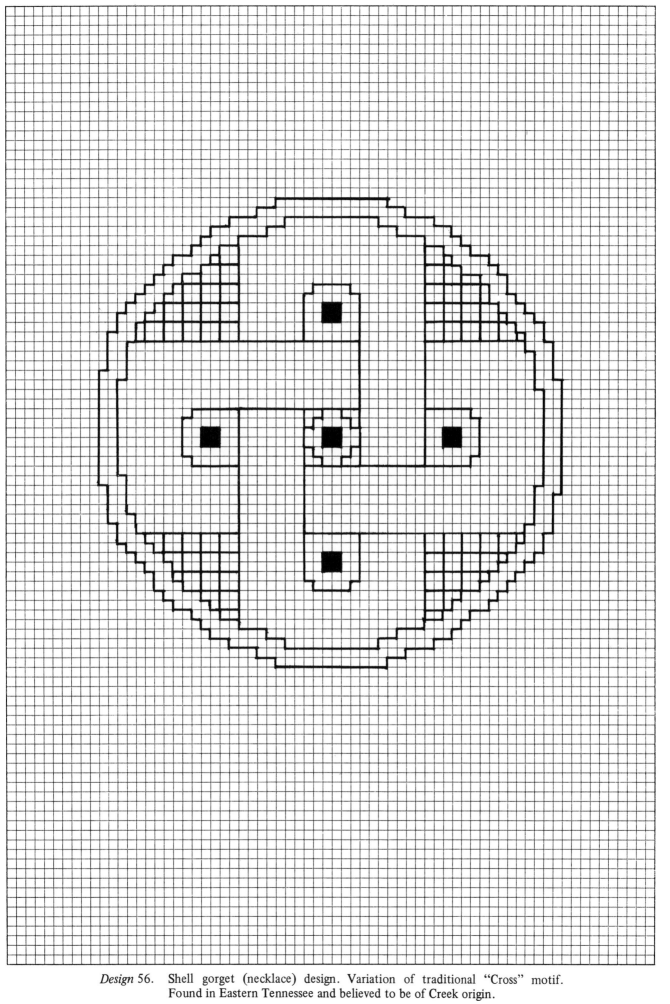

*Design* 56.  Shell gorget (necklace) design. Variation of traditional "Cross" motif.
Found in Eastern Tennessee and believed to be of Creek origin.

*Design* 57.   Stone earspool (primitive earrings) design. Found at Moundville, Alabama excavation and believed to be of Creek origin.

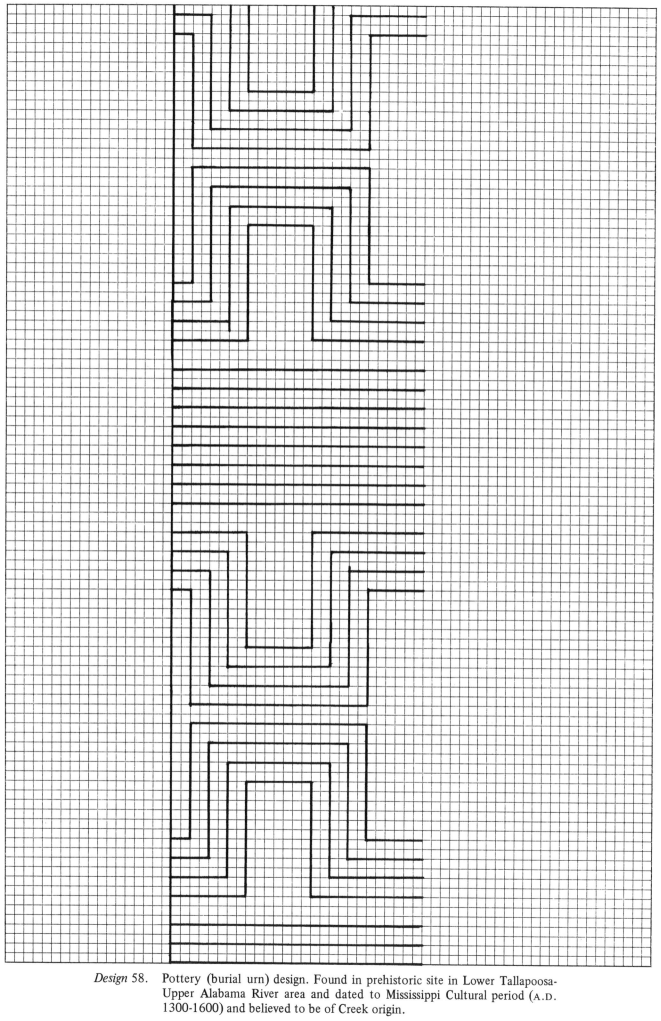

*Design* 58.    Pottery (burial urn) design. Found in prehistoric site in Lower Tallapoosa-
Upper Alabama River area and dated to Mississippi Cultural period (A.D.
1300-1600) and believed to be of Creek origin.

*Design* 59.   Shell gorget (necklace) design. Shows two woodpeckers within the traditional symbol for the "World" — a cross within a circle. Prehistoric and found at the Etowah site excavation in Georgia, it is believed to be of Creek origin.

*Design* 60. Design on a ceremonial stone disk found at Moundville, Alabama, excavation and believed to be of Creek origin.

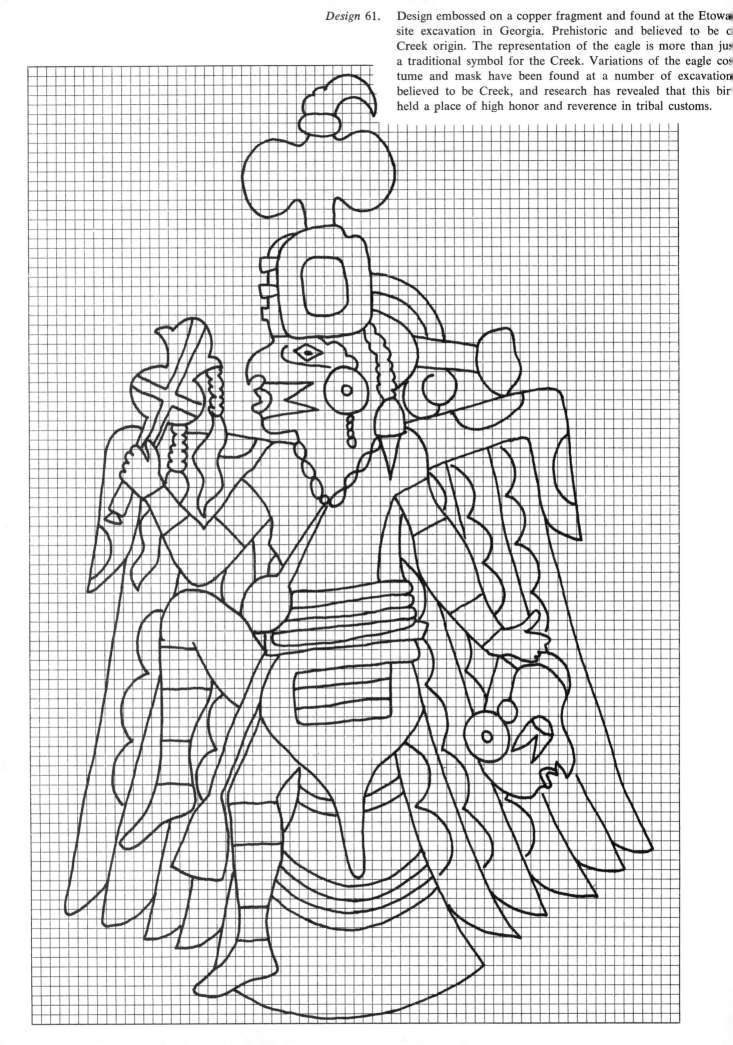

*Design* 61. Design embossed on a copper fragment and found at the Etowa[h] site excavation in Georgia. Prehistoric and believed to be [of] Creek origin. The representation of the eagle is more than jus[t] a traditional symbol for the Creek. Variations of the eagle co[s]tume and mask have been found at a number of excavation[s] believed to be Creek, and research has revealed that this bir[d] held a place of high honor and reverence in tribal customs.

*Design* 62.    Same design as the preceding page. This is a further adaptation of the design for needlepoint.

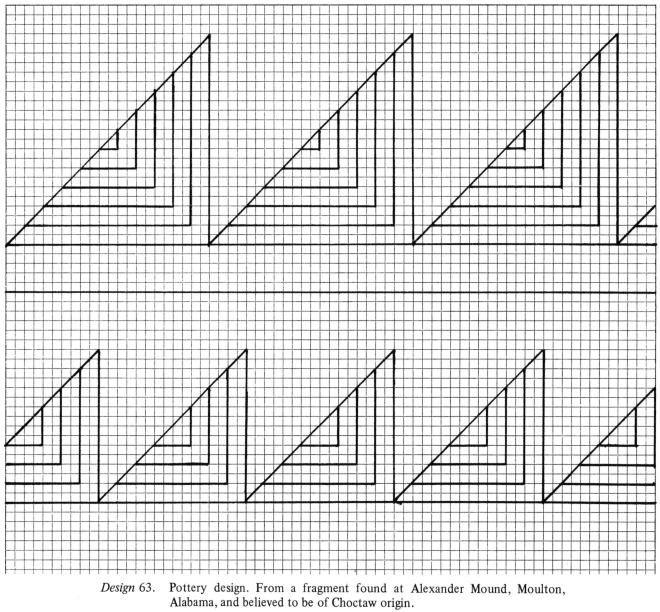

*Design* 63.   Pottery design. From a fragment found at Alexander Mound, Moulton, Alabama, and believed to be of Choctaw origin.

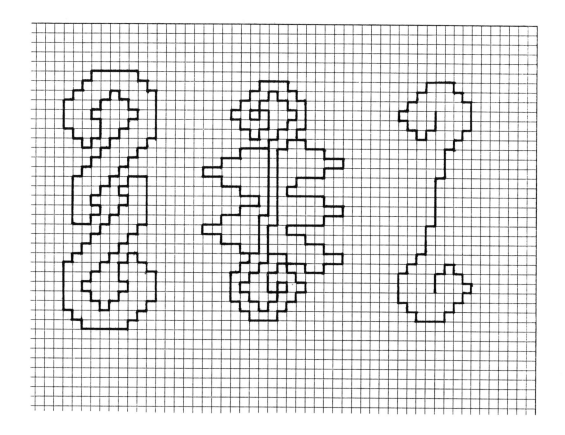

*Design* 64.  (Top) Variations of a design of a lacrosse players belt.
(Bottom) Belt design of a lacrosse player.

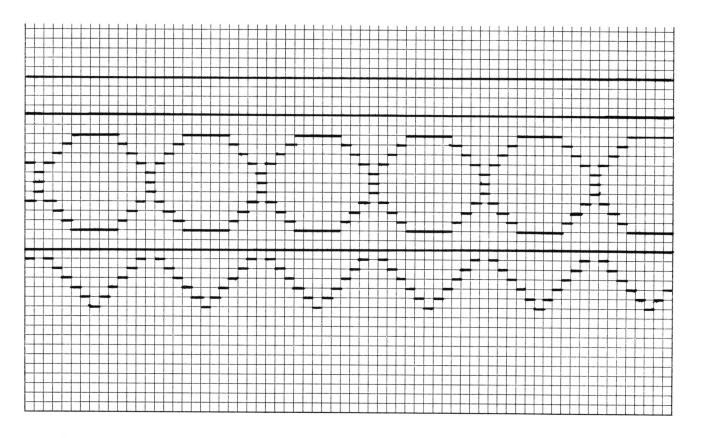

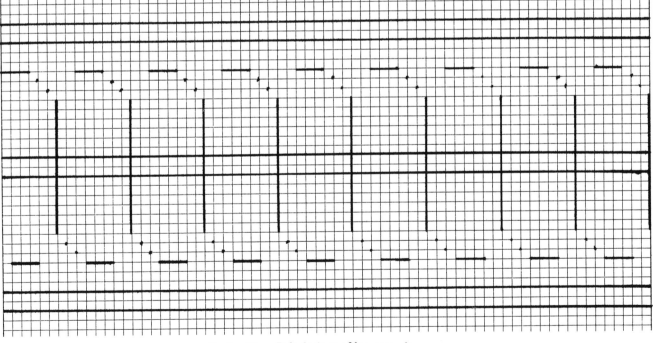

*Design* 65.   Belt designs of lacrosse players.

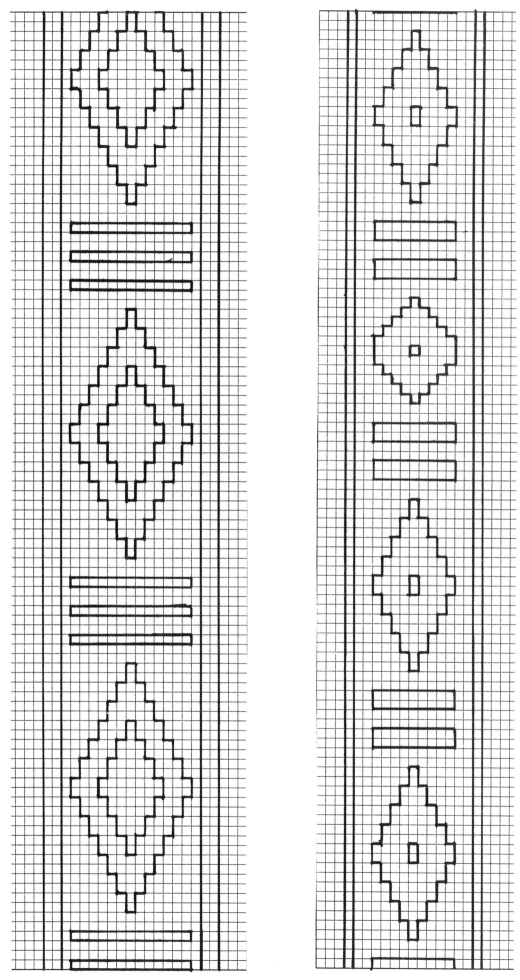

*Design* 66.   Belt designs of lacrosse players.

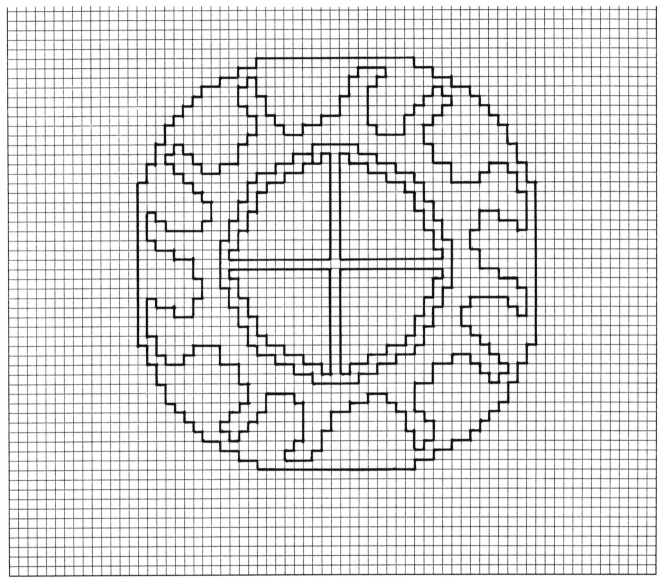

*Design* 67.   Beaded sash design.

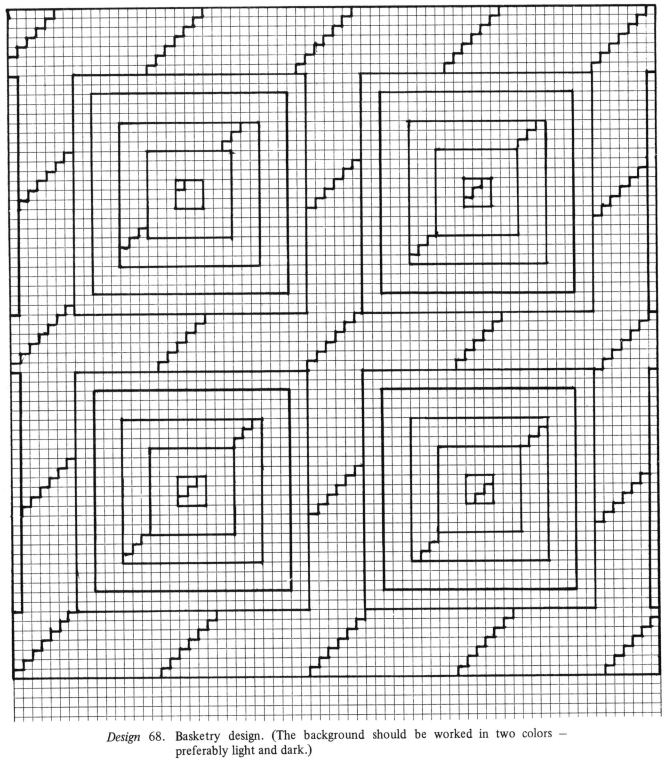

*Design* 68. Basketry design. (The background should be worked in two colors —
preferably light and dark.)

*Design 69.* Basketry design.

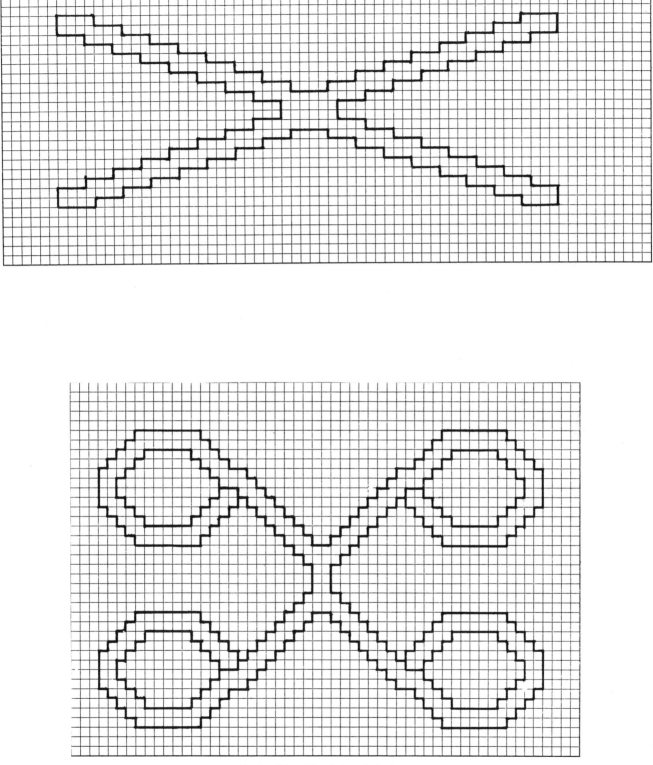

*Design* 70.   Beaded sash designs.

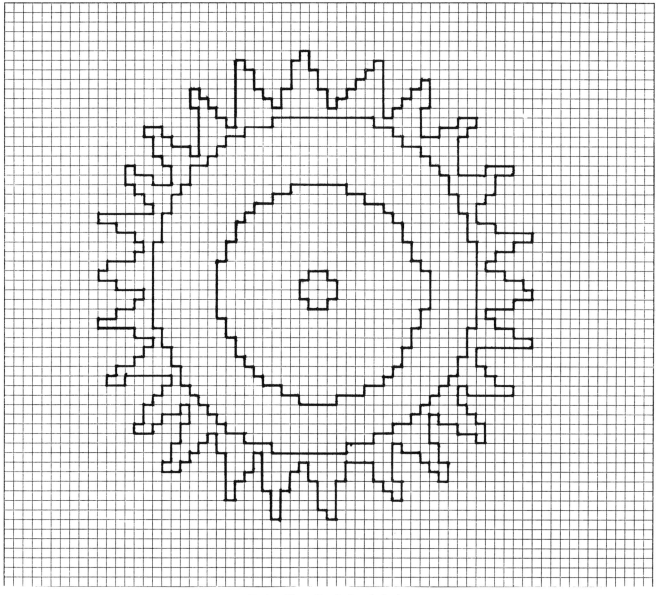

*Design* 71.   Beaded sash design.

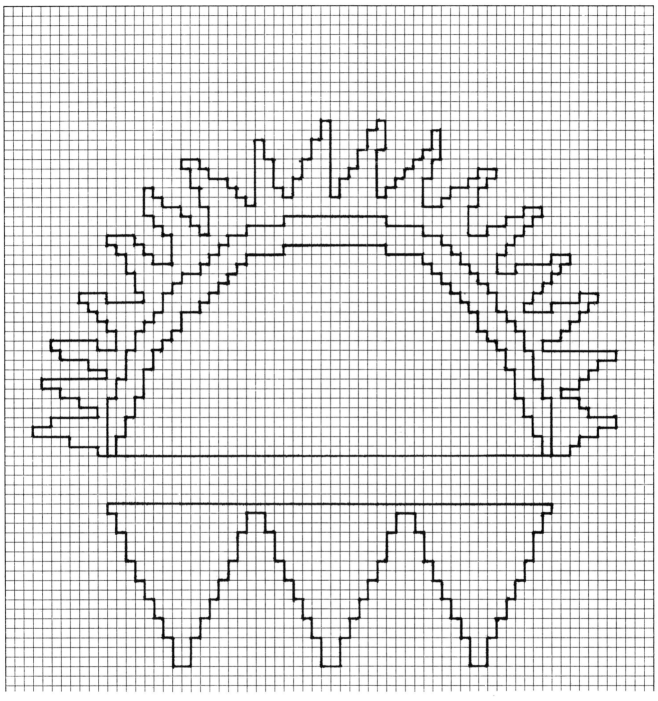

*Design* 72.   Beaded sash design.

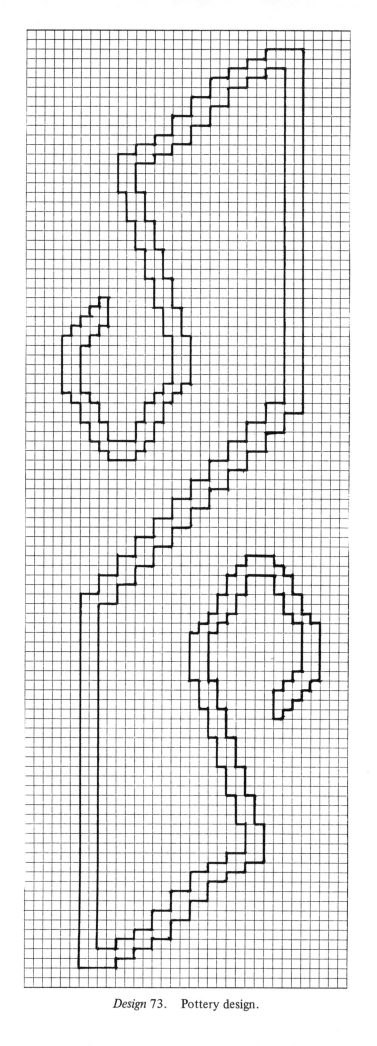

*Design* 73.   Pottery design.

*Design* 74.   Beaded sash design.

*Design* 75.　Basketry design. (The background should be worked in two colors — preferably light and dark.)

*Design* 76.   Beaded sash design.

*Design* 77.   Basketry design.

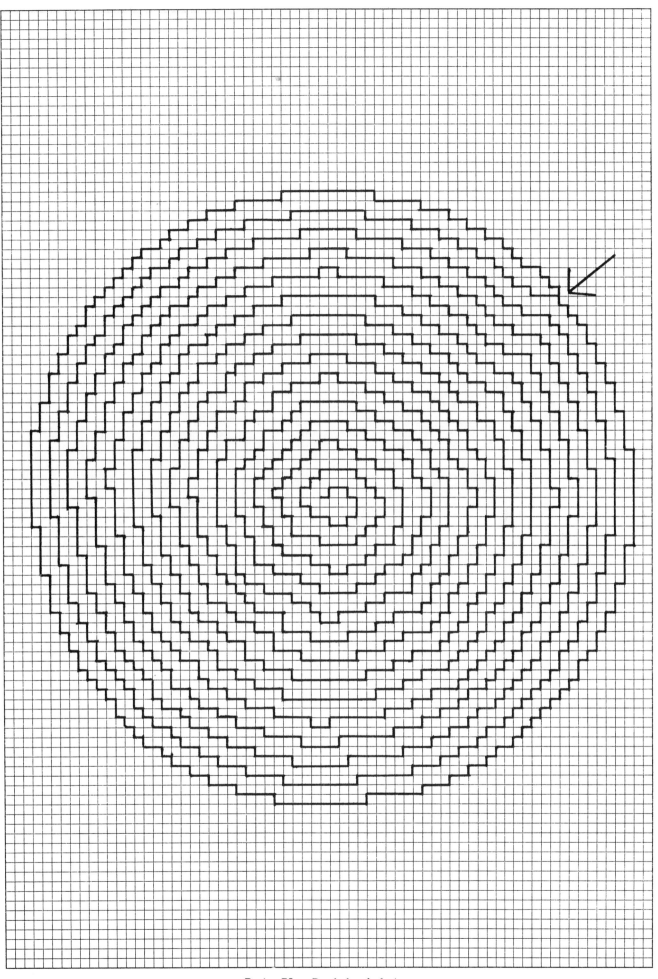

*Design* 78.   Beaded sash design.

*Design* 79.    Stone disk found in Mississippi and believed to be of Choctaw origin.

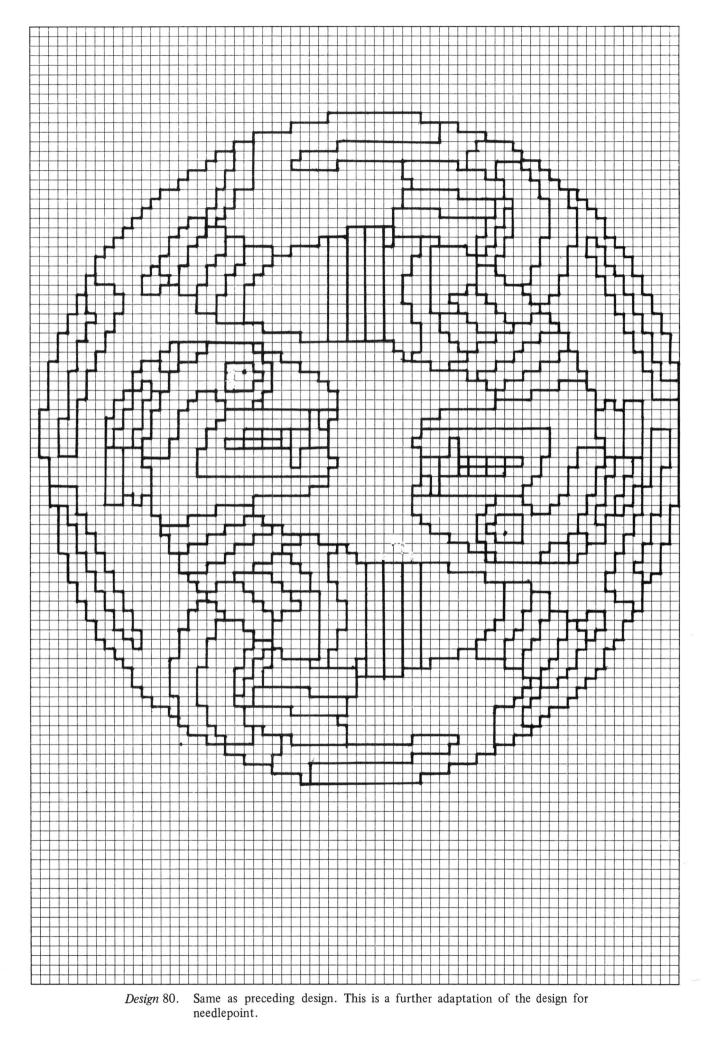

*Design* 80.  Same as preceding design. This is a further adaptation of the design for needlepoint.

**Seminole**

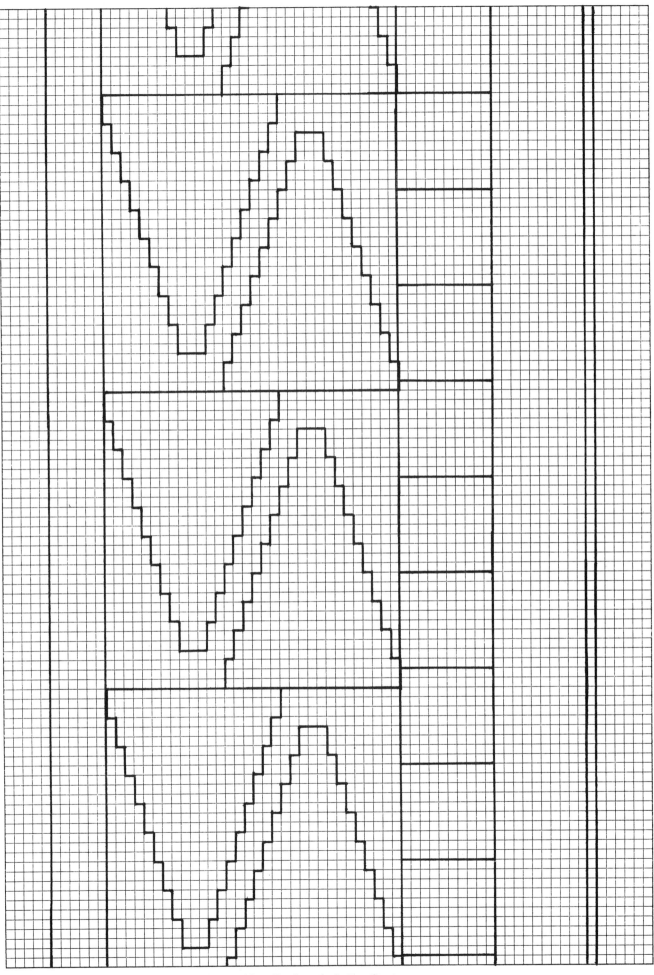

*Design* 81.   Patchwork design. Contemporary.

*Design* 82.   Patchwork design. Contemporary.

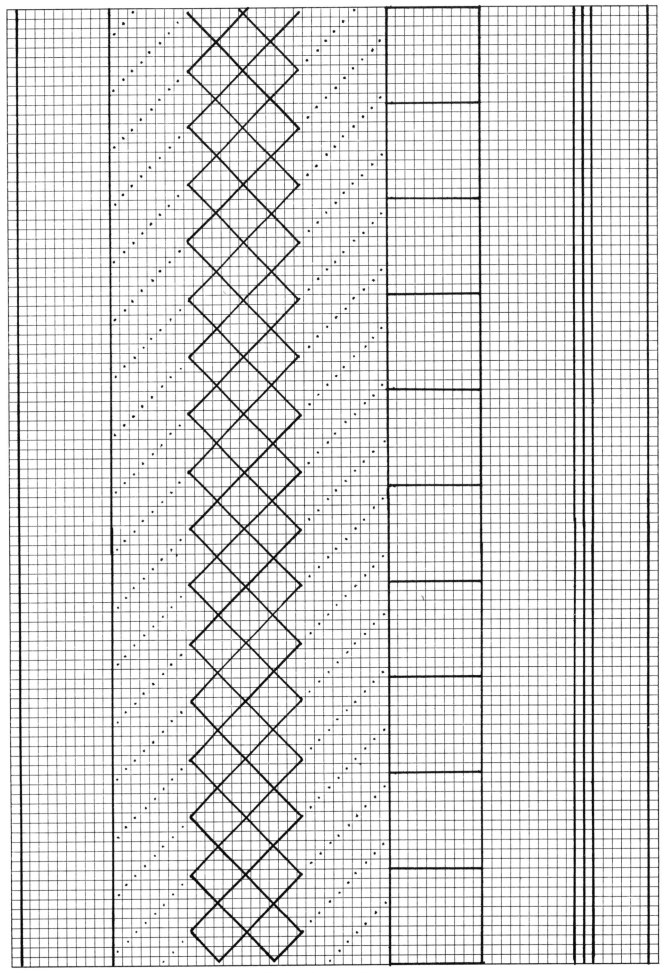

*Design* 83. Patchwork design. Contemporary.

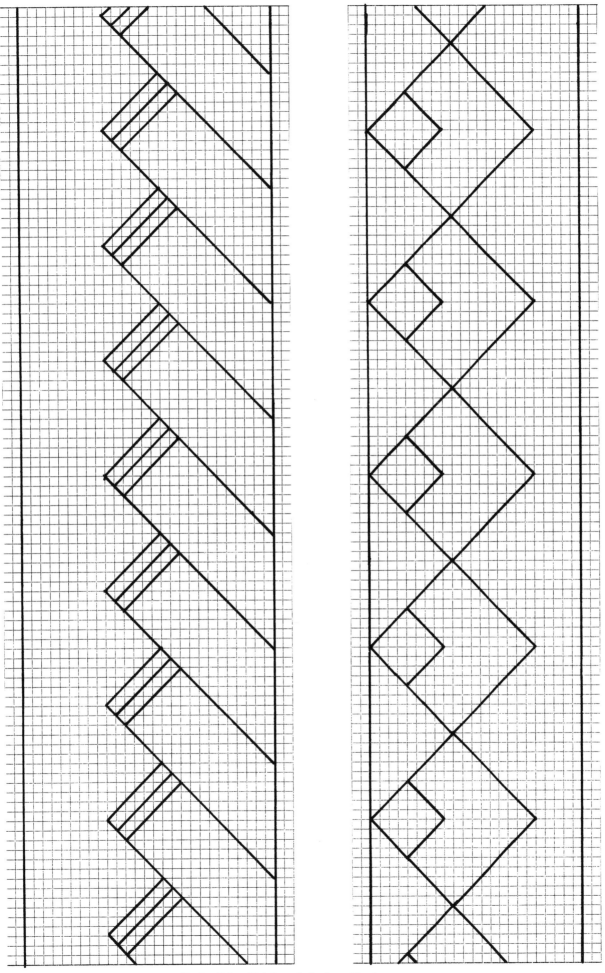

*Design* 84.   Patchwork designs. Contemporary.

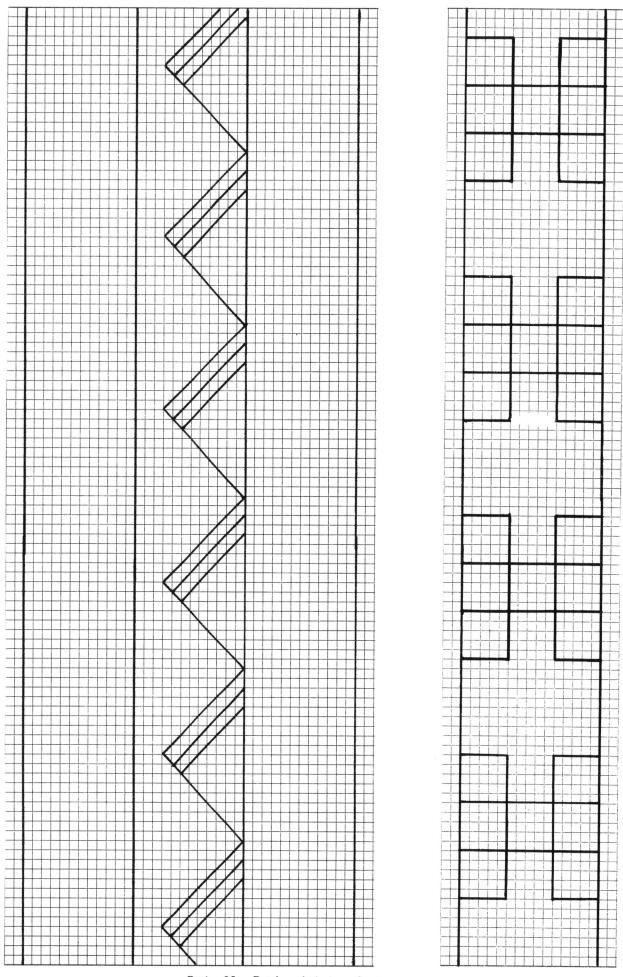

*Design* 85. Patchwork designs. Contemporary.

*Design* 86.    Patchwork design. Around 1910.

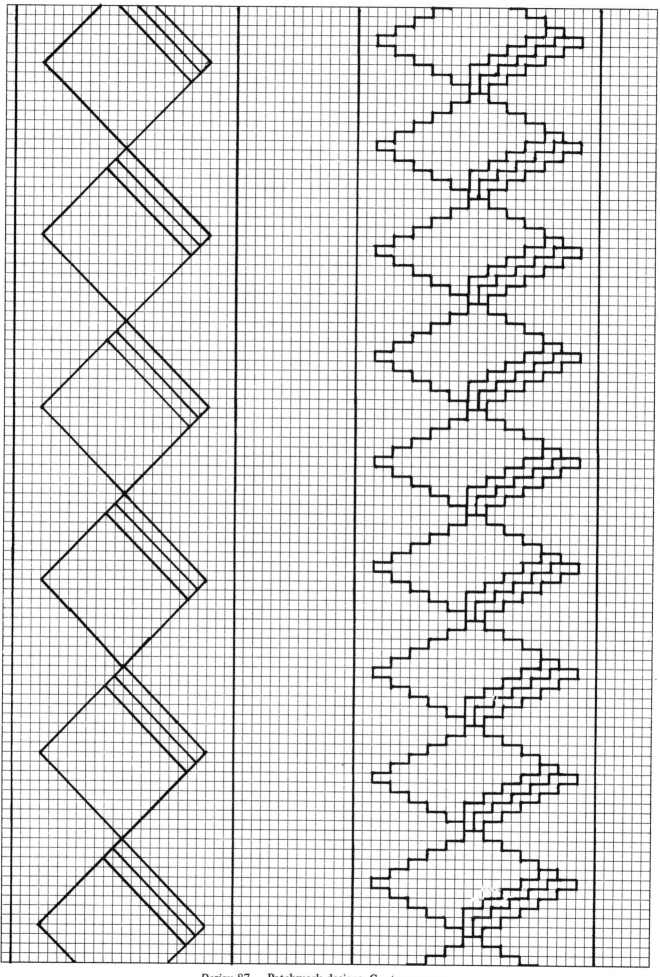

*Design* 87. Patchwork designs. Contemporary.

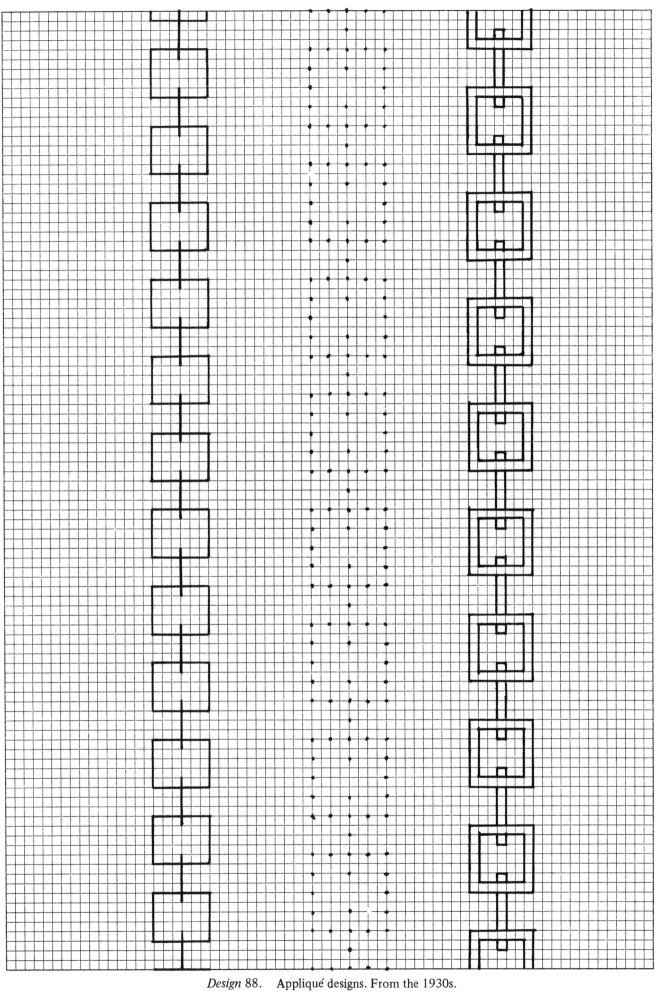

*Design* 88.    Appliqué designs. From the 1930s.

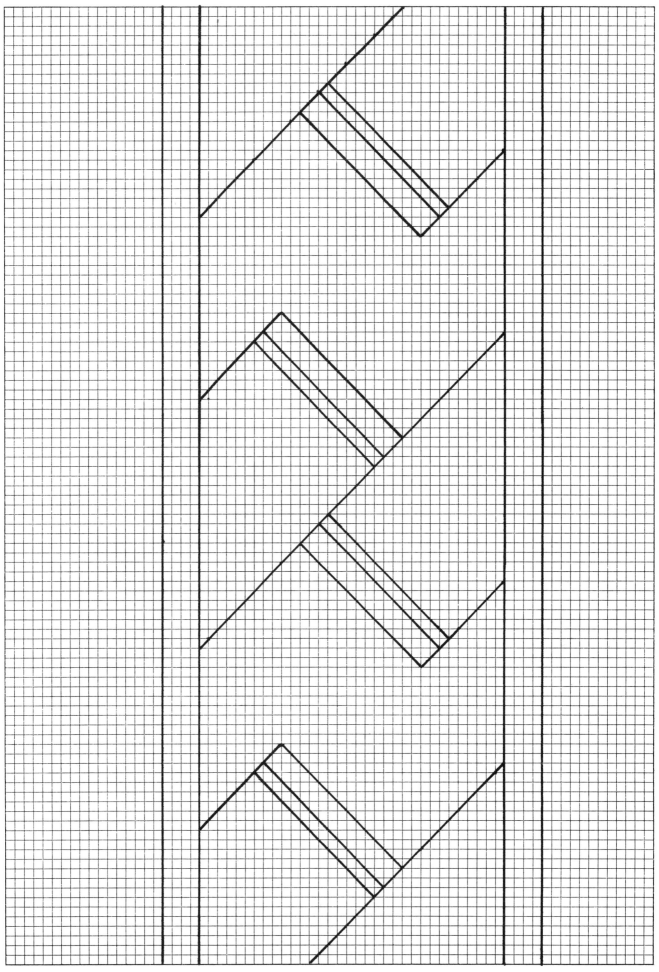

*Design* 89. Patchwork design. Contemporary.

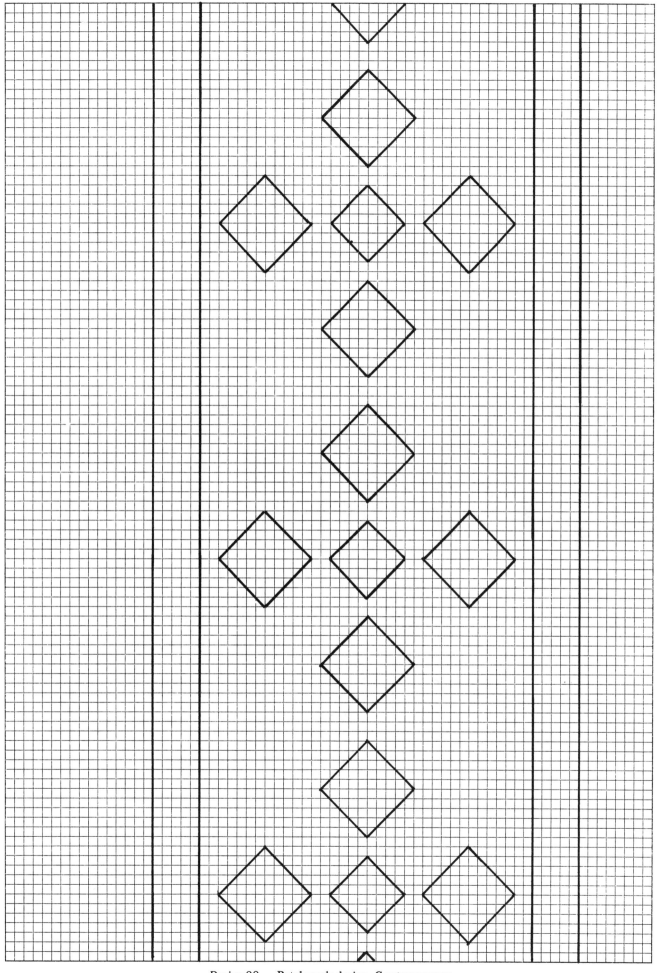

*Design* 90.    Patchwork design. Contemporary.

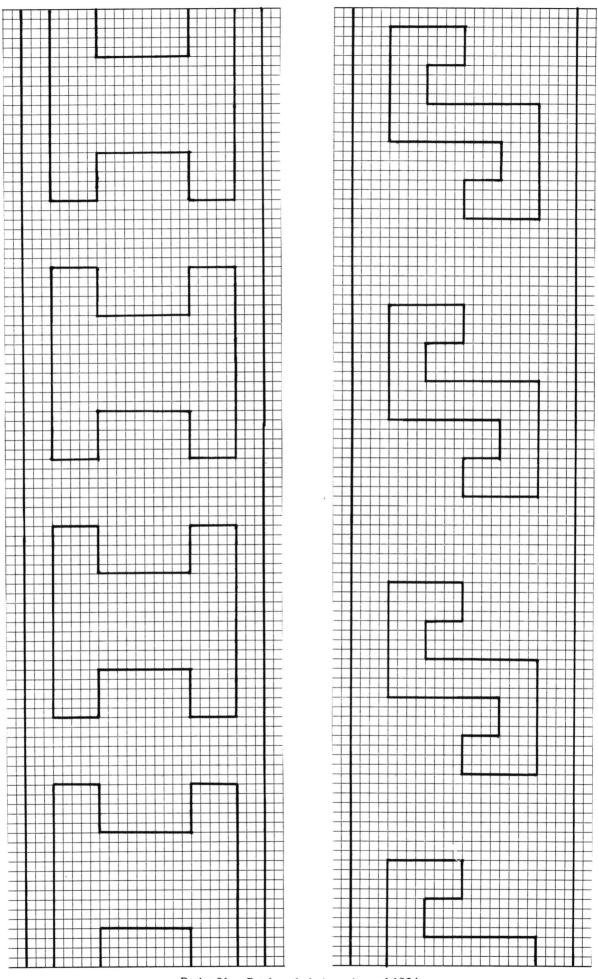

*Design* 91.    Patchwork designs. Around 1934.

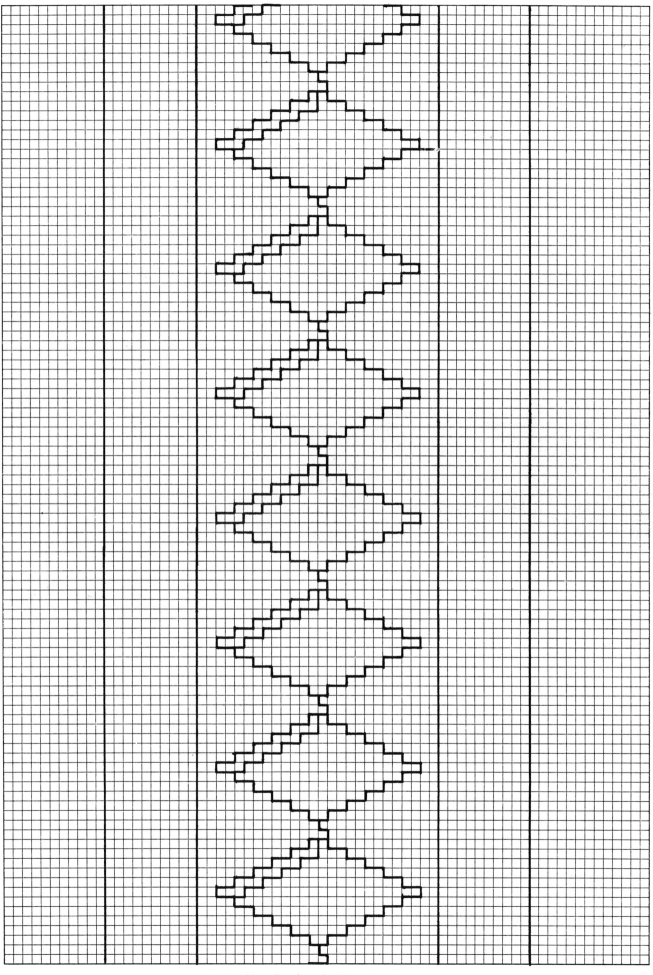

*Design* 92.   Patchwork design. Contemporary.

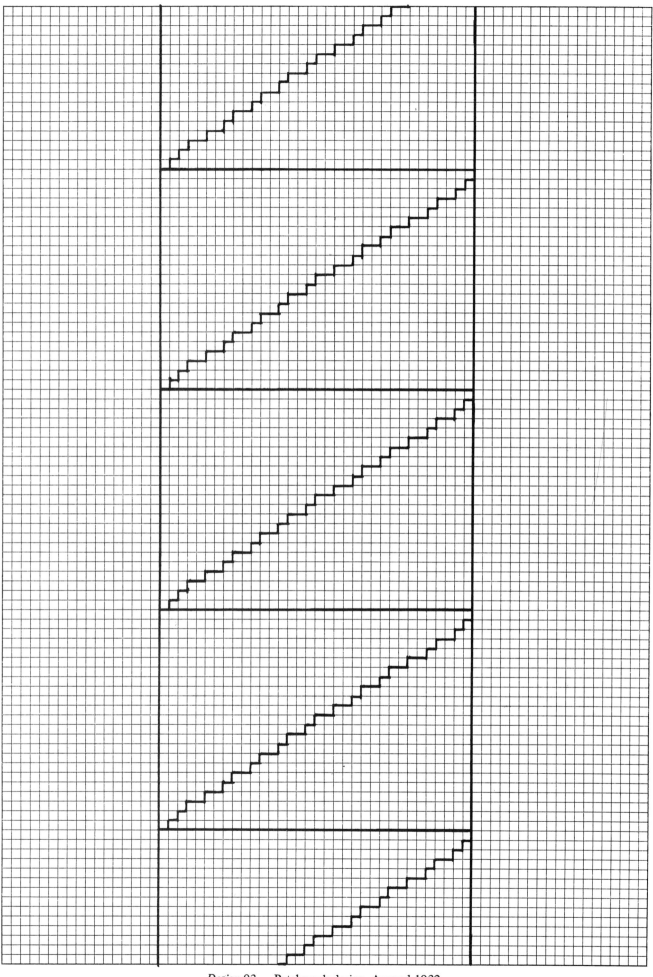

*Design* 93.   Patchwork design. Around 1932.

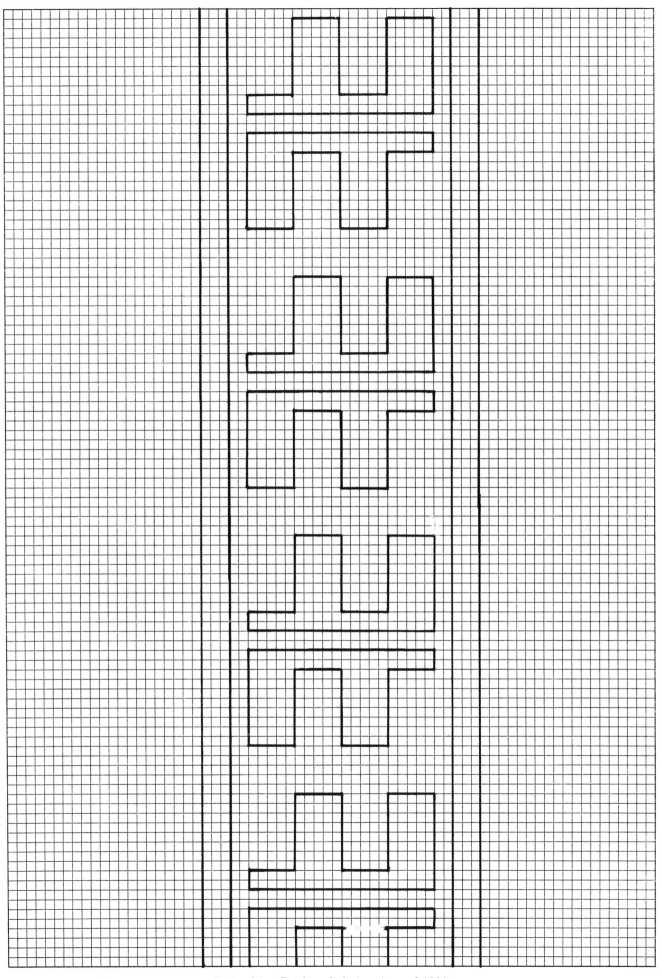

*Design* 94.   Patchwork design. Around 1890.

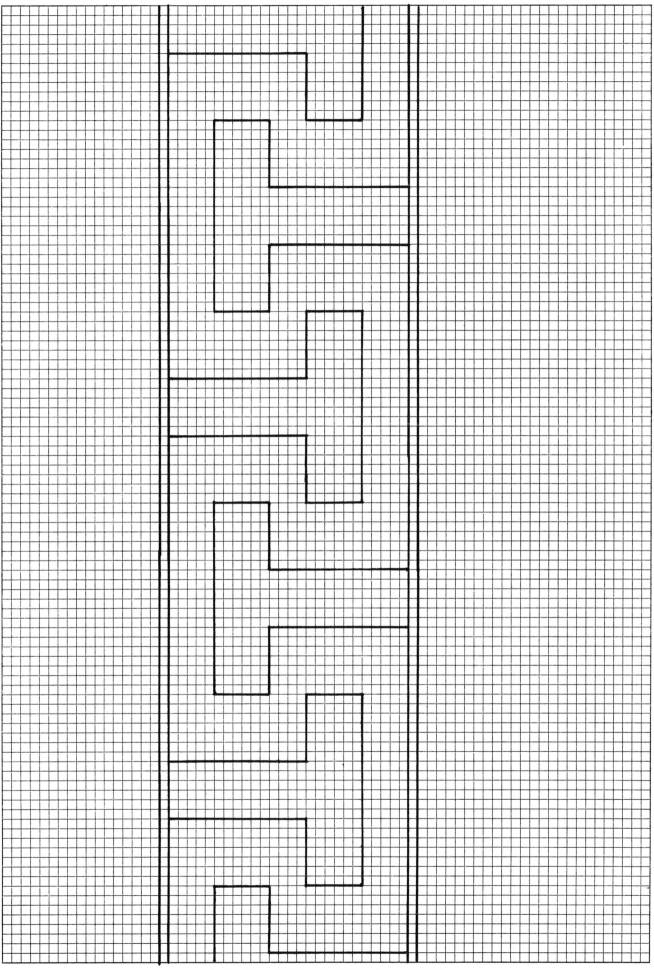

*Design* 95.    Patchwork design. Around 1934.

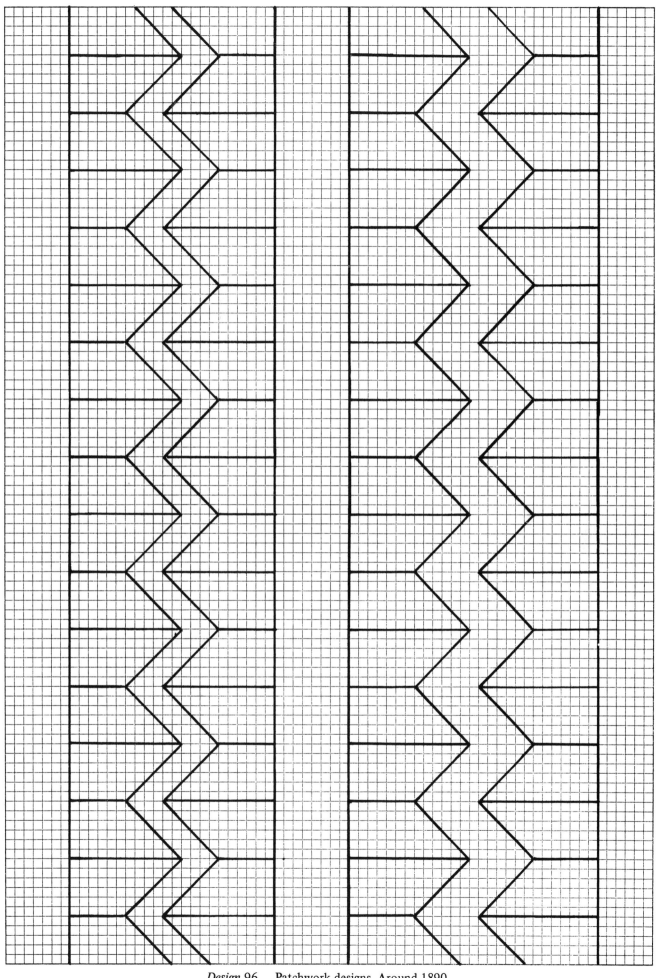

*Design* 96.   Patchwork designs. Around 1890.

*Design* 97.   Textile design. Contemporary.

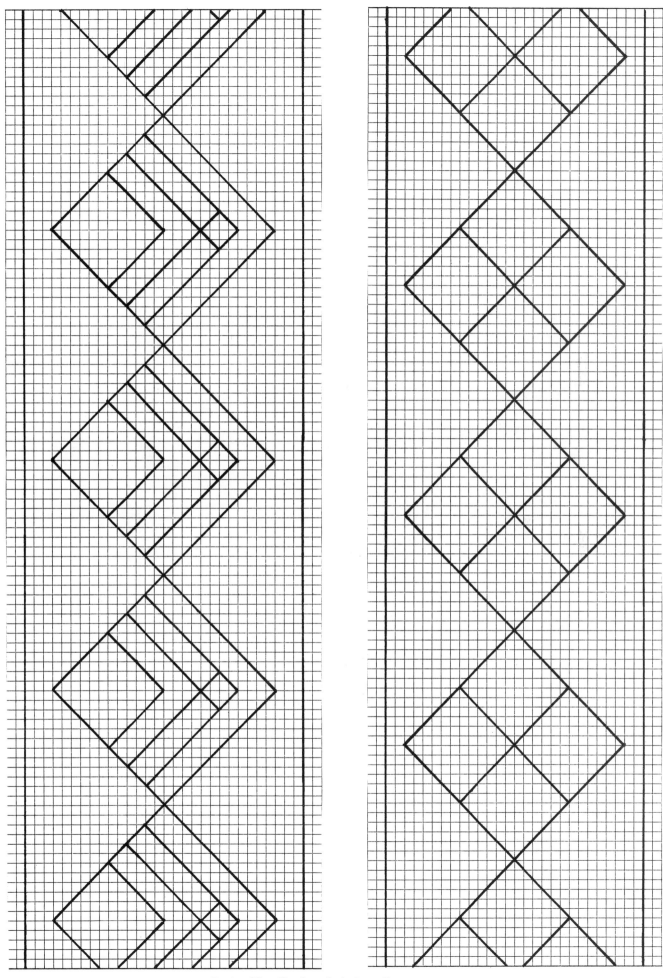

*Design* 98.   Patchwork designs. Contemporary.

*Design* 99.    Finger-woven belt designs. Around 1910. A technique used as opposed to
loom weaving.

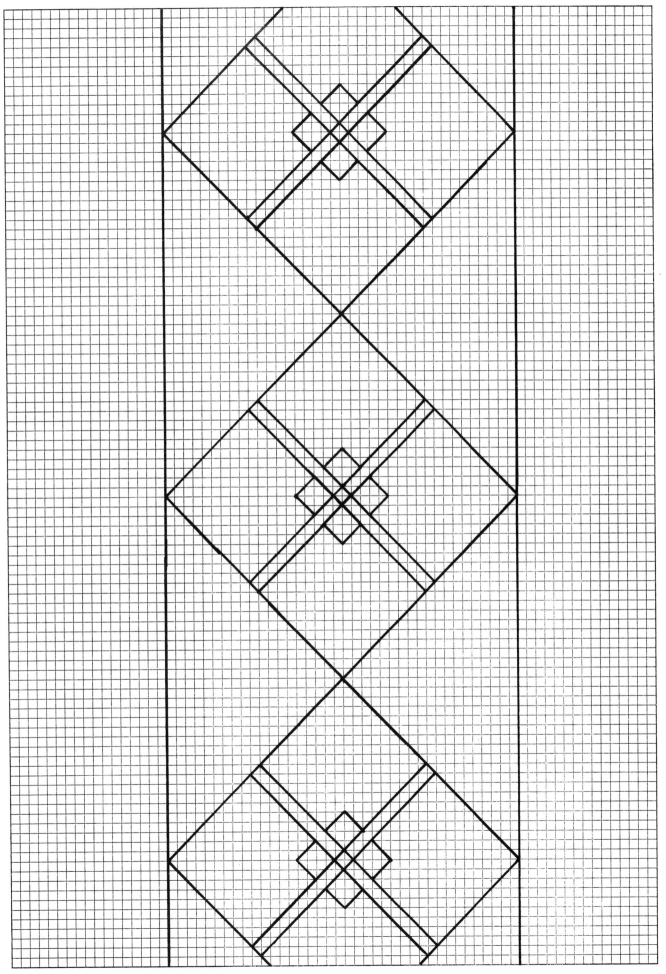

*Design* 100. Patchwork design. Contemporary.

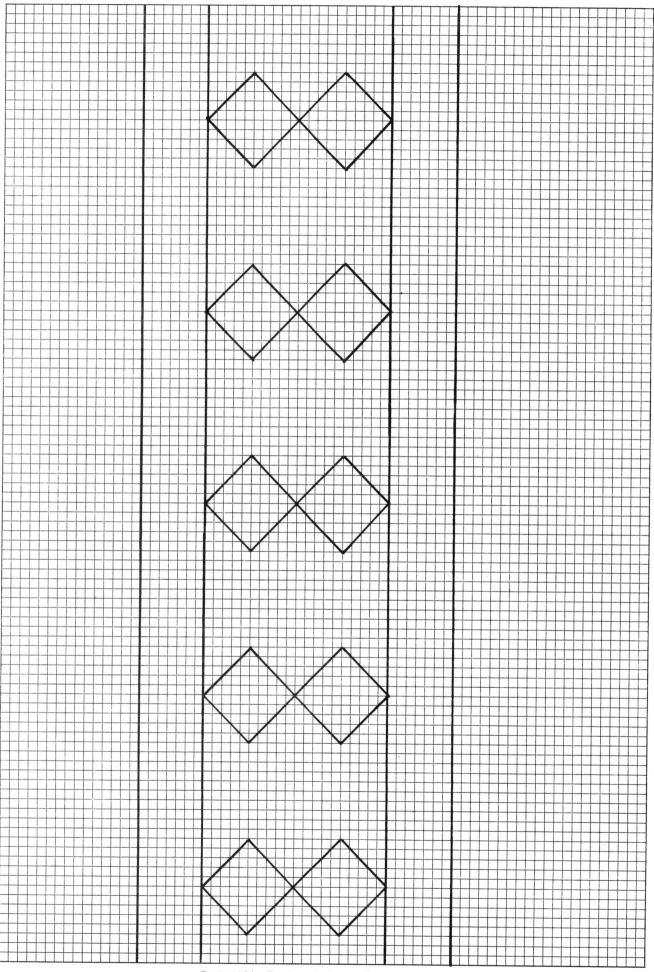

*Design* 101. Patchwork design. Contemporary.

*Design* 102. Patchwork design. Contemporary.

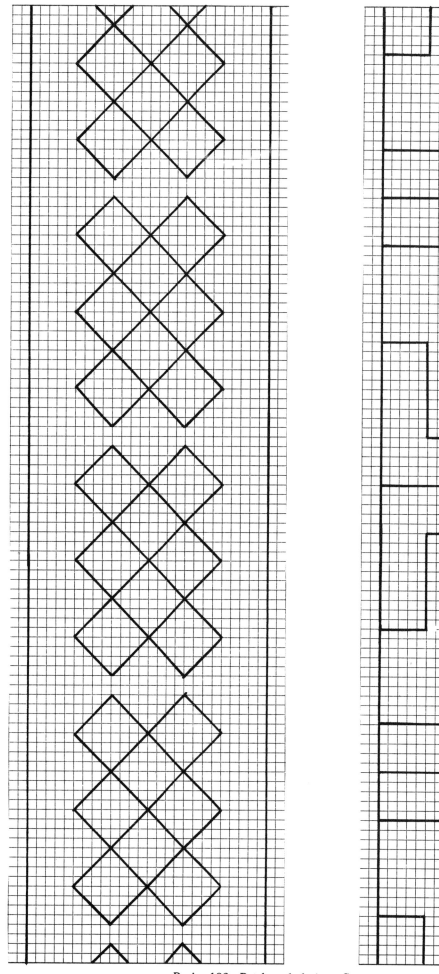
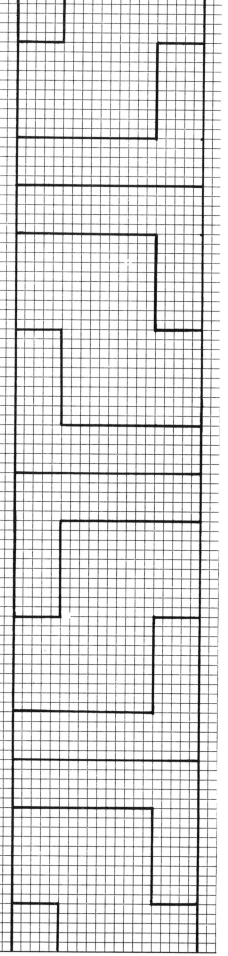

*Design* 103. Patchwork designs. Contemporary.

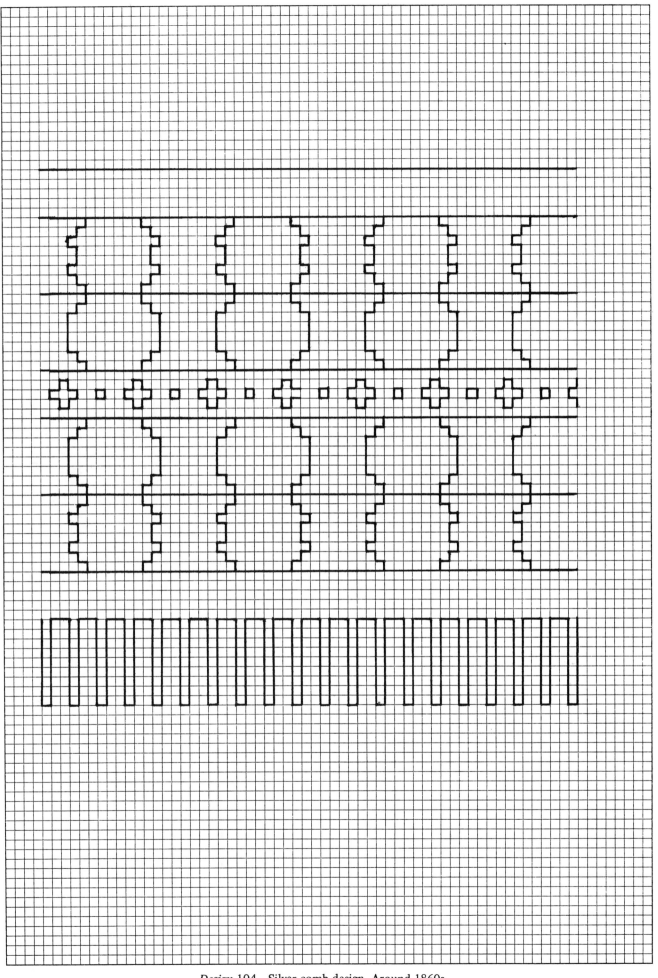

*Design* 104.  Silver comb design. Around 1860s.

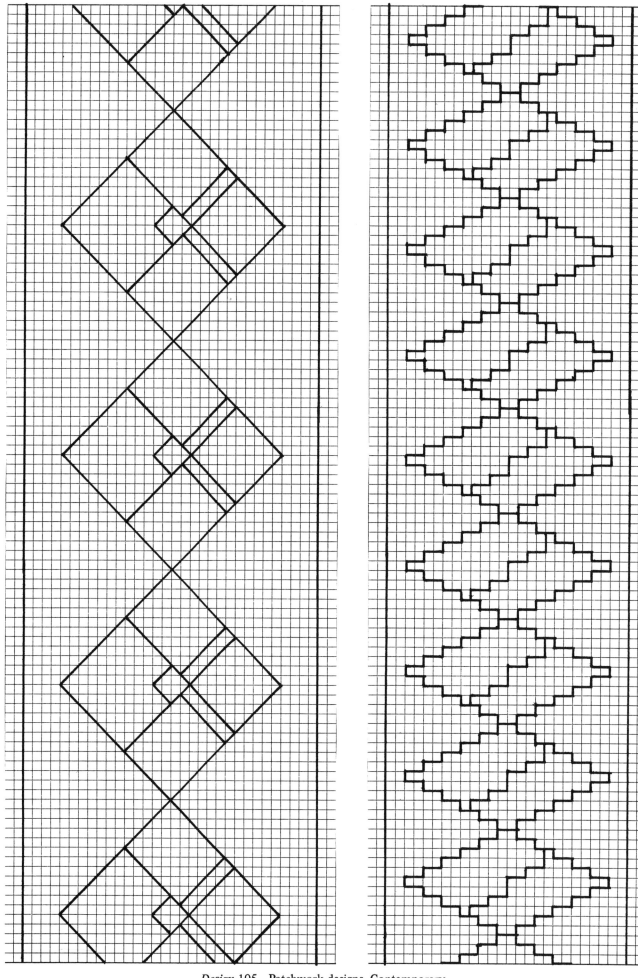

*Design* 105. Patchwork designs. Contemporary.

*Design* 106. Patchwork design. Contemporary.

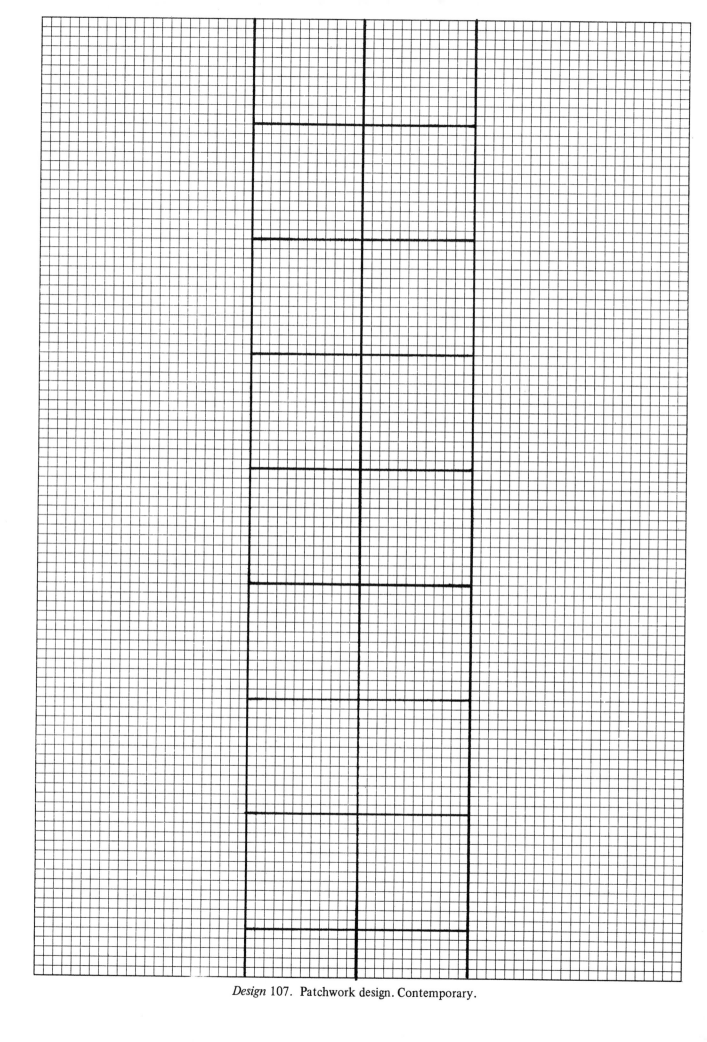

*Design* 107. Patchwork design. Contemporary.

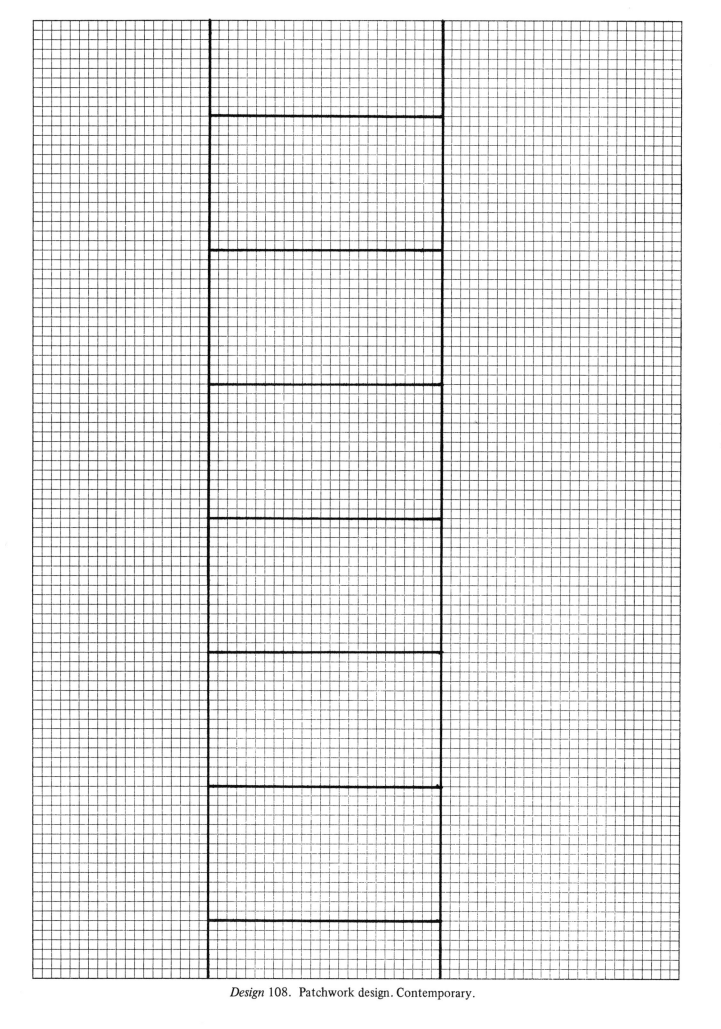

*Design* 108.  Patchwork design. Contemporary.

*Design* 109. Beaded sash designs. Around 1930s.

*Design* 110. Woven scarf design. Around 1930s.

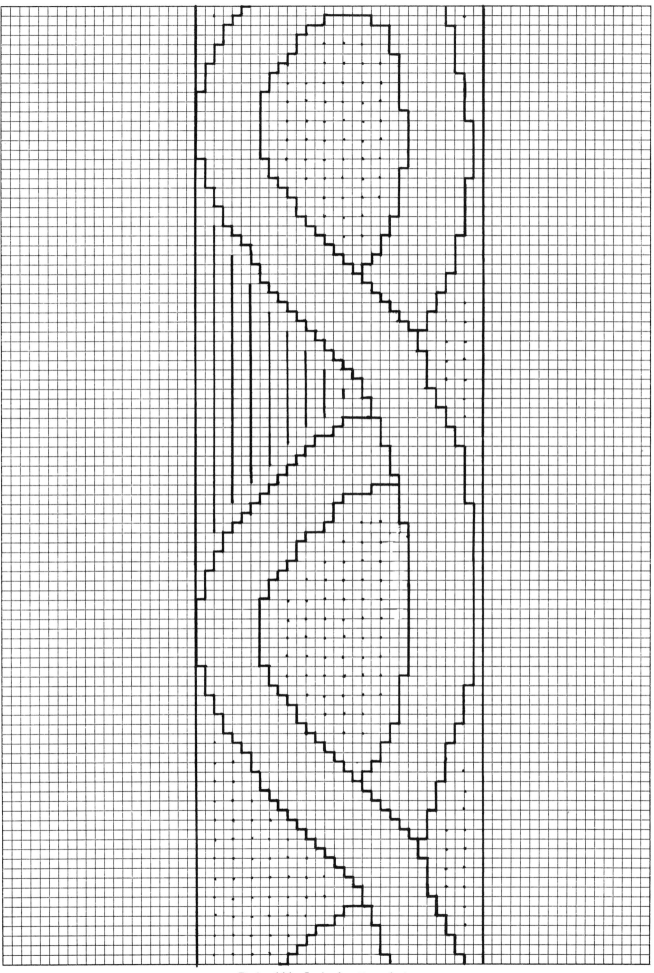

*Design* 111. Incised pottery design.

*Design* 112. Beaded sash design. Owned by Billy Bowlegs, famous Seminole warrior in the Second Seminole War, 1836-42. His Indian name was Halpattermicco and he was made a principal chief during their struggle to resist removal to the reservations of the Oklahoma territory.

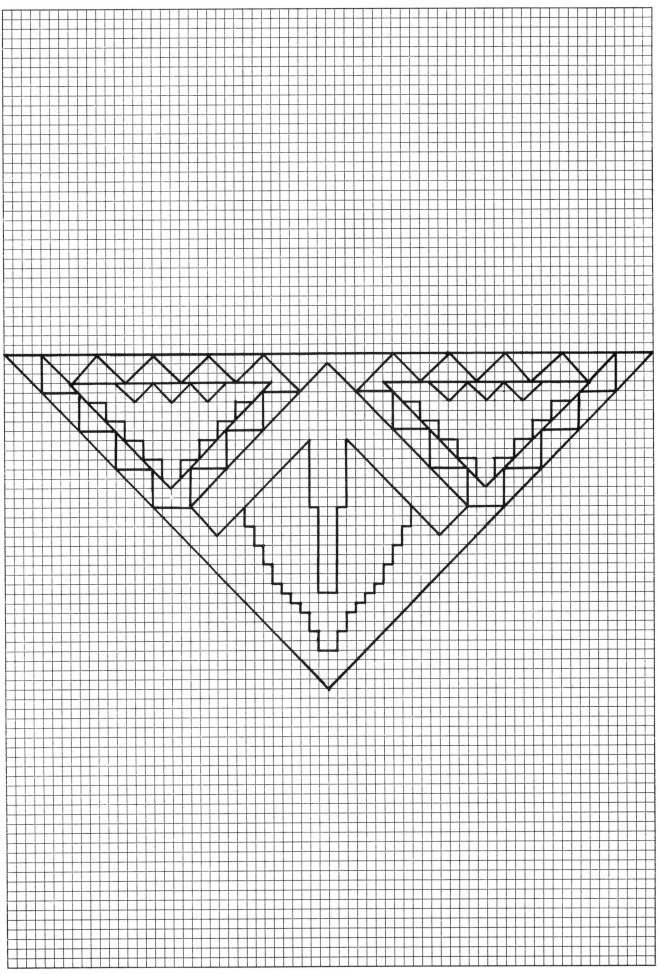

*Design* 113. Beaded shoulder bag design.

*Design* 114. Beaded shoulder bag design.

*Design* 115. Combination silver neck and head band design. Around the 1700s and from adornment worn by Mico Chlucco, the Long Warrior who was probably one of the earliest kings of the Seminoles.

*Design* 116.  Silver ornament designs.  Around 1900.

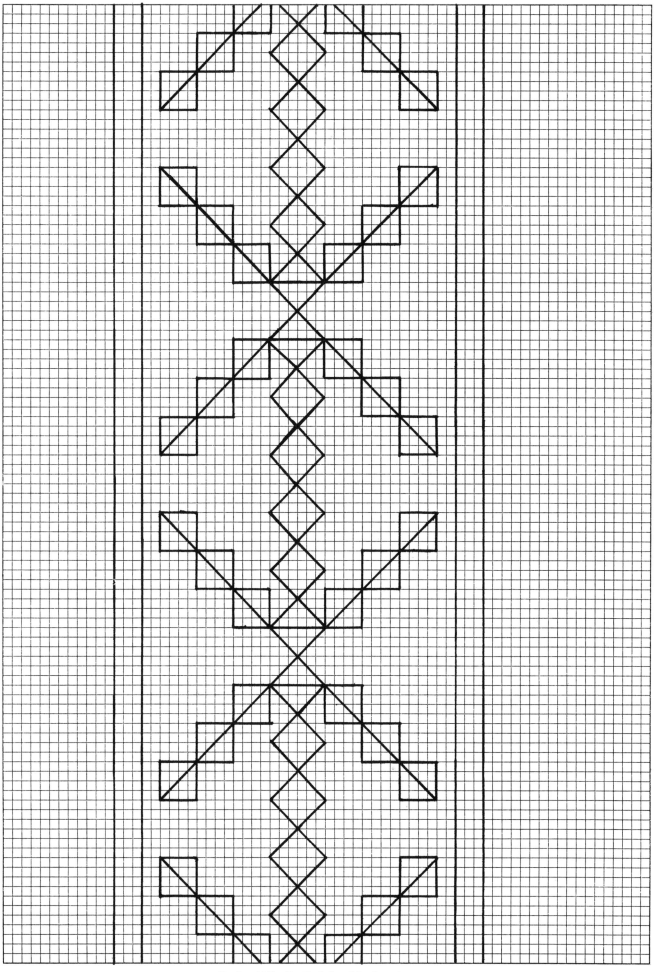

*Design* 117.  Beaded shoulder bag design.

*Design* 118. Beaded sash design.

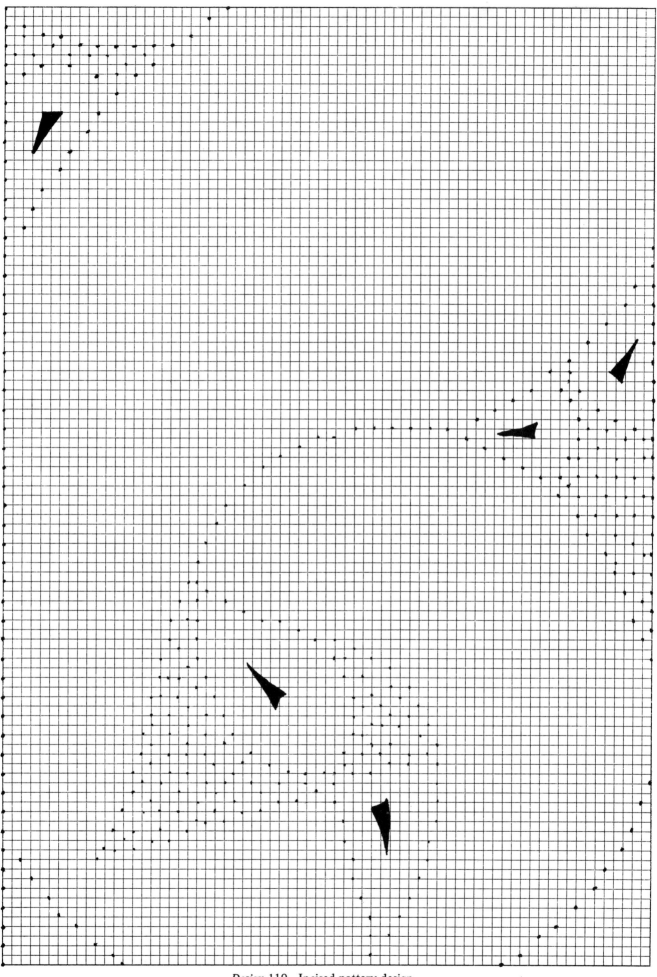

*Design* 119. Incised pottery design.

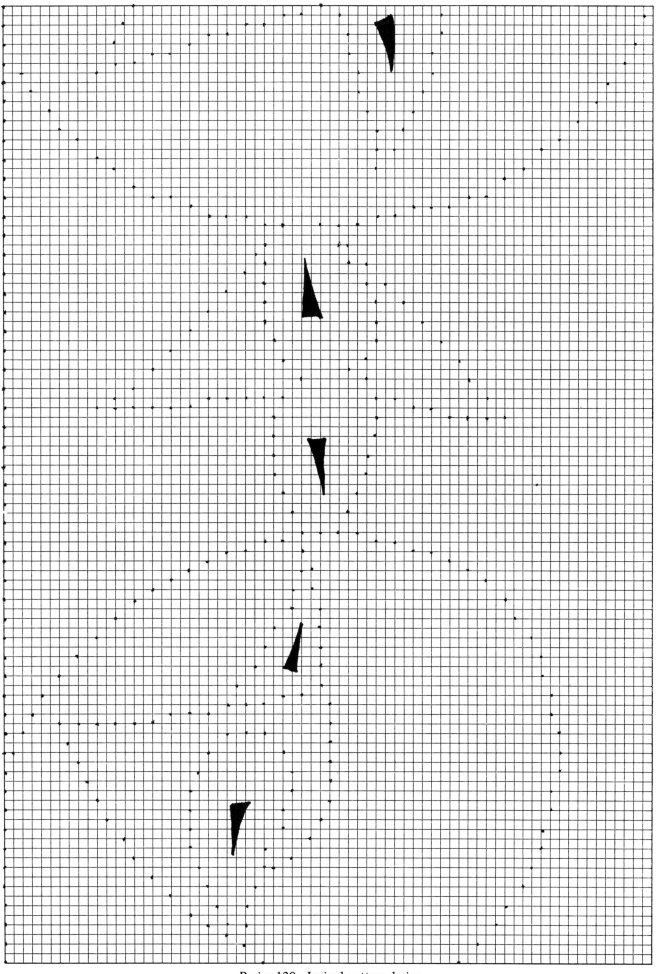

*Design* 120.  Incised pottery design.

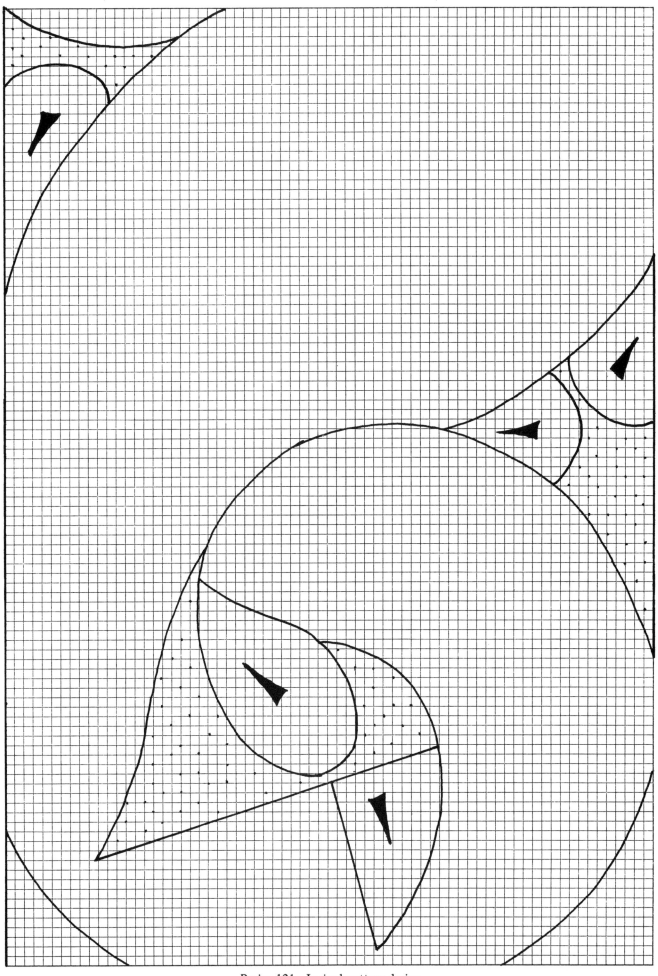

*Design* 121. Incised pottery design.

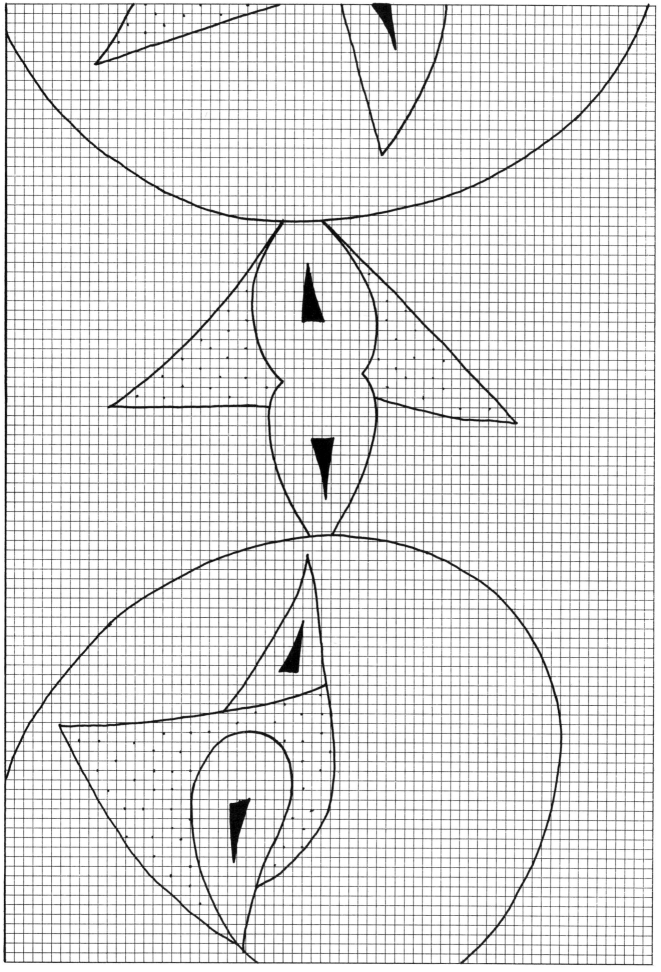

*Design* 122. Incised pottery design.

# INDEX

Accessories, 29

Blocking, 36
Brick stitch: left-handed, 26; right-handed, 21

Canvas, 29-32
Cherokee: designs, 38-67; history, 13
Cherokee diamond stitch: left-handed, 27; right-handed, 22
Chickasaw, 15
Choctaw: designs, 100-117; history, 15
Continental stitch: left-handed, 23; right-handed, 19
Creek: designs, 68-99; history, 14

Designs: Cherokee, 38-67; Choctaw, 100-117; Creek, 68-99; Seminole, 118-59

Finishing, 35

Flame stitch: left-handed, 27; right-handed, 21

Needles, 29

St. Andrew cross stitch: left-handed, 25; right-handed, 20
St. George cross stitch: left-handed, 25; right-handed, 20
Seminole: designs, 118-59; history, 16
Slanting gobelin stitch: left-handed, 23; right-handed, 19
Stitches, 18-27
Straight gobelin stitch: left-handed, 25; right-handed, 20

Transferring designs, 33

Working tips and habits, 33

Yarn, 28